Geology and Ore Deposits of the Takilma and Waldo Mining Districts of Josephine County, Oregon

U.S. Geological Survey Bulletin 846-B

Compiled by the staff of
the United States Geological Survey

with an introduction by Kerby Jackson

Introduction

It has been over eighty years since the United States Geological Survey released its Bulletin 846-B, which was otherwise known as "Geology and ore deposits of the Takilma-Waldo district, Oregon (including the Blue Creek district)".

The author of this work, the late Philip J. Shenon, was well known in those days for his expertise and his written contributions to the mining industry, especially in Idaho, Oregon and Montana.

In addition to this volume, some of Shenon's written contributions to the mining industry include such titles as "Geology of the Robertson, Humdinger, and Robert E. Gold mines, southwestern Oregon" (1933), "Copper Deposits in the Squaw Creek and Silver Peak Districts and at the Almeda Mine, Southwestern Oregon" (1933), "Geology of Elk City Mining District, Idaho" (1934), "Geology and Ore Occurances of the Hog Heaven Mining District, Flathead County, Montana" (1936), "Geology and ore deposits near Edwardsburg and Thunder mountain, Idaho" (1936), "Geology and ore deposits near Murray, Idaho" (1938), "The silver belt of the Coeur d'Alene district, Idaho" (1939), "Geologic study of the Big Deer Creek area, Blackbird Mining District, Lemhi County, Idaho" (1956), "Guidebook of the Geology of the Coeur D'Alane Mining Discrict" (1961), "Economic Geology: Today and Tomorrow" (1962) and others. Shenon also wrote many articles including "Down Idaho's River of No Return" about a geological expedition he led down the Salmon River which was published by National Geographic Magazine in July of 1936.

Shenon, a one time President of the Society of Economic Geologists, also left behind a legacy of paperwork about his geological work spanning from 1926 to 1973 which is housed at the Harold B. Lee Library at Brigham Young University in Provo, Utah. The collection consists of three boxes of "*notebooks, note cards, (a) typewritten thesis, travel logs, budget books, geological maps, and photographs. The materials relate to Shenon's work as a geologist ... in the American West and his teaching career*", according to the curator.

Though he studied the economic geology of Southern Oregon only a few short years, Philip Shenon was quick to establish himself as an expert on mining in Southern Oregon and his written works remain important to miners and historians in the area. His research and field studies on local copper mines, in particular, remain of extreme historical importance in large part because Shenon was able to tour such mines as the Almeda, Cowboy, Queen of Bronze and others in the late 1920's and early 1930's, interviewing their operators, just before the runs of these mines came to their final close. Indeed, much of what we do know about the Queen of Bronze Mine, as well as the Cowboy and other copper mines in Josephine County's Takilma Mining District, is due mostly to the written records left behind by Philip J. Shenon.

In addition to his work on the copper mines in the vicinity of Takilma, Shenon also left us with an important record of some of the oldest placer mines in the Waldo District, which in fact, include some of the very earliest of Oregon's gold mines. When Shenon came upon the local scene in the late 1920's, while these mines had already been operating for decades, the mine operators that Shenon interviewed were only one and two generations removed from the first miners who came into the area in the early 1850's, some of them being the children and grandchildren of the first men who into the area looking for gold.

It has often been said that *"gold is where you find it"*, but even beginning prospectors understand that their chances for finding something of value in the earth or in the streams of the Golden West are dramatically increased by going back to those places where gold and other minerals were once mined by our forerunners. Despite this, much of the contemporary information on local mining history that is currently available is mostly a result of mere local folklore and persistent rumors of major strikes, the details and facts of which, have long been distorted. Long gone are the old timers and with them, the days of first hand knowledge of the mines of the area and how they operated. Also long gone are most of their notes, their assay reports, their mine maps and personal scrapbooks, along with most of the surveys and reports that were performed for them by private and government geologists. Even published books such as this one are often retired to the local landfill or backyard burn pile by the descendents of those old timers and disappear at an alarming rate. Despite the fact that we live in the so-called "Information Age" where information is supposedly only the push of a button on a keyboard away, true insight into mining properties remains illusive and hard to come by, even to those of us who seek out this sort of information as if our lives depend upon it. Without this type of information readily available to the average independent miner, there is little hope that our metal mining industry will ever recover.

This important volume and others like it, are being presented in their entirety again, in the hope that the average prospector will no longer stumble through the overgrown hills and the tailing strewn creeks without being well informed enough to have a chance to succeed at his ventures.

Kerby Jackson
Josephine County, Oregon
February 2014

GEOLOGY AND ORE DEPOSITS OF THE TAKILMA-WALDO DISTRICT, OREGON, INCLUDING THE BLUE CREEK DISTRICT

By Philip J. Shenon

ABSTRACT

Two areas and their included mineral deposits, situated in Josephine County, southwestern Oregon, are described in this report. They lie within the Klamath Mountains, a region which is made up for the most part of rugged ridges trending in various directions but which, when viewed from higher summits, resembles a dissected plateau and is known as the Klamath peneplain. Rocks of both igneous and sedimentary origin are abundant in the districts described. The marine sedimentary rocks of the areas comprise a thick series of Carboniferous strata, with some interbedded volcanic rocks, and portions of the Galice formation, of Jurassic age, and of an Upper Cretaceous formation. The rocks of fluviatile origin include Tertiary conglomerate, Pleistocene valley fill, termed the "Llano de Oro formation," and somewhat later Pleistocene gravel and alluvium, in part glacial debris. Recent gravel is found along the present streams. The igneous rocks include several varieties of greenstone of probable Paleozoic and Mesozoic age and serpentine of late Jurassic or early Cretaceous age.

The Takilma-Waldo area contains some of the most productive mines in Oregon. Copper deposits have produced ore worth $1,700,000; and gold and platinum, with a minimum value of $4,000,000, have been mined from placers. Chromite deposits have produced a small tonnage. Four of the described copper deposits are in formations designated greenstone; one is in serpentine. All, however, are close to greenstone-serpentine contacts, and, as indicated by their mineral constitution, all are high-temperature deposits. The ore occurs as disconnected bodies of irregular outline ranging in size from mere stringers to deposits containing as much as 10,000 tons. In greenstone the ore minerals do not form solid bodies within the limits of the deposits but are interspersed with bands and irregular areas of altered rock, much after the manner of mineralized shear zones; in serpentine they occur as "boulderlike" bodies surrounded by slickensided wall rock. Oxidation and enrichment extend only to shallow depths. The prevailing hypogene ore minerals are pyrite, chalcopyrite, cubanite, pyrrhotite, sphalerite, and, in the serpentine deposit, cobaltite. Minerals of later origin are iron oxides, malachite, azurite, cuprite, chrysocolla, and chalcocite. The gangue in the greenstone deposits is principally altered wall rock, quartz, and calcite. Calcite has been introduced later than the sulphides and in some places in sufficient amounts to deplete the grade of the ore. Serpentine, calcite, epidote, and quartz are the abundant gangue minerals in the serpentine deposit. Two very large leached outcrops occur in greenstone at the Turner (Albright) mine. Pyrite and some chalcopyrite have been found in tunnels beneath the leached outcrops, but sufficient prospecting has not been done to determine the worth of the property. No ore is known to have been shipped from the vicinity.

Three kinds of placer deposits occur in the Takilma-Waldo district—Tertiary conglomerate, residual surface mantle derived from the weathering of the Tertiary conglomerate, and re-sorted deposits in gulches or on flats on or below areas of Tertiary conglomerate. The re-sorted deposits are the most valuable, although considerable gold and platinum have been mined from deposits of the other kinds. The gold and platinum in the Tertiary conglomerate are distributed more or less evenly throughout the formation, but in the re-sorted placers they occur principally on or near bedrock. Field relations indicate that the Tertiary conglomerate is a stream deposit and that most of the gold and platinum in the other placers of the district have been derived from it. Field evidence also indicates that most of the gold and platinum in the Tertiary conglomerate were transported in the boulders, to be later liberated by weathering. Furthermore, transportation of portions of the weathered conglomerate by rain wash and streams caused concentration of the gold and platinum along bedrock in the re-sorted placers.

The copper mines of the Takilma-Waldo district appear capable of producing a considerable future tonnage but, because of the high mining and transportation costs, can be worked at a profit only when the demand for copper is strong. Very little ore is blocked out in any of the mines and yet some can be seen in most of the stopes. Largely because of the irregular outline of the deposits and because the ore shoots terminate abruptly, mining and development proceed more or less simultaneously, and hence very little ore is proved in advance of mining. Because of the nature of the deposits, projections of ore bodies as a basis for tonnage estimates of ore in place are extremely hazardous and not reliable. New ore bodies will most likely be found even below the deepest level if the price of copper should rise sufficiently to justify prospecting.

Because of its low gold content, it is doubtful whether the Tertiary conglomerate should be classified among the reserve placer deposits. If, however, at some future time exceptionally low costs should prevail, much of the formation might prove workable. The existing remnants of the formation contain many million cubic yards of material. The re-sorted (Llano de Oro) gravel is to be regarded as the chief source of future placer production. Information given by reports of trustworthy engineers and from other reliable sources indicates that areas of this formation aggregating several hundred acres contain enough gold to be profitably mined. In addition, there is much ground that is probably gold bearing, and the areas of known and probable value together aggregate at least a thousand acres. The deposit ranges from a few feet to 80 feet or more in depth, and its volume probably equals or exceeds that of the Tertiary formation. In the prospected areas, information indicates that the gold content ranges generally from 10 to 60 cents a cubic yard, with streaks that are much richer.

INTRODUCTION

Field work and acknowledgments.—This report presents the results of investigations undertaken as a part of the cooperative survey of the mineral resources of Oregon by the State Mining Board and the United States Geological Survey. The field work upon which the report is based was done under the supervision of J. T. Pardee, during July and part of August, 1930, by P. J. Shenon, assisted by Duncan Johnson and Aubrey Walker. This party mapped in detail over 33 square miles in the Takilma-Waldo district and a small area in the Blue Creek district and investigated the copper mines of both districts and the placer and chromite deposits near Takilma and Waldo.

It is a pleasure to acknowledge the courtesies extended by all with whom the writer came into contact during the course of the field work. Messrs. E. H. Messenger, C. D. Cameron, Kameel Khoeery, and J. L. Eggers, of Takilma; G. M. Esterly, of Waldo; and Edward Turner, of O'Brien, all gave information and generous amounts of their time. Others, too numerous to mention, cooperated in many ways.

Location and transportation facilities.—So far as mining is concerned, the districts described in this report are located in one of the most productive parts of Oregon. Notable amounts of copper have come from the mines near Takilma, and the placers near Waldo have been and still are the largest producers of gold and platinum in the State.

The portions of southwestern Oregon described are in Josephine County, just north of the California boundary. (See pl. 9.) The main line of the Southern Pacific Railroad runs through Grants Pass, 40 to 50 miles northeast of the mapped areas, and a branch line has been extended southward from that city, so that Takilma is now only 27 miles from a shipping point. Excellent paved highways serve as main arteries for truck haulage and automobile travel throughout most of southwestern Oregon, and Crescent City, a California seaport, appears to offer an outlet for some of the products of the region.

Literature and maps.—The areas covered in this report are included in the Kerby quadrangle, a topographic map of which has been published by the United States Geological Survey on a scale of 1 to 125,000, or about 2 miles to the inch, with a contour interval of 100 feet. The following list includes the principal papers bearing directly on the geology and ore deposits of the described areas:

Browne, J. R., Report upon the mineral resources of the States and Territories west of the Rocky Mountains for 1866, p. 141, 1867; idem for 1867, pp. 576–596, 1868.

Diller, J. S., Mineral resources of southwestern Oregon: U. S. Geol. Survey Bull. 546, 1914.

Diller, J. S., and Kay, G. F., Mineral resources of the Grants Pass quadrangle and bordering districts, Oreg.: U. S. Geol. Survey Bull. 380, pp. 48–79, 1909.

Hornor, R. R., Notes on the black-sand deposits of southern Oregon and northern California: U. S. Bur. Mines Tech. Paper 196, pp. 29–33, 1918.

Kellogg, A. E., Placer mining in Oregon: Eng. and Min. Jour., vol. 108, pp. 90–91, 1919.

Kellogg, A. E., Auriferous gravels of southwest Oregon: Min. Jour., Phoenix, Ariz., vol. 11, No. 20, pp. 3–6, 54–55, March 15, 1928.

Kellogg, A. E., Origin of the copper in southwestern Oregon: Min. Jour., Phoenix, Ariz., vol. 12, No. 2, pp. 9–10, June, 1928.

Kellogg, A. E., Llano de Oro placers, Waldo district, Oreg.: Min. Jour., Phoenix, Ariz., vol. 12, No 4, pp. 9–11, July 15, 1928.

Nicol, J. M., Placers of Waldo, southern Oregon: Min. and Sci. Press, vol. 99, pp. 122–124, July 24, 1909.

Parks, H. M., and Swartley, A. M., Handbook of the mining industry of Oregon—alphabetical list of properties; description of mining districts: Mineral Resources of Oregon, vol. 2, No. 4, Oregon Bur. Mines and Geology, 1916.

Smith, W. D., and Packard, E. L., The salient features of the geology of Oregon: Oregon Univ. Bull., vol. 16, No. 7, July, 1919.

Winchell, A. N., Petrology and mineral resources of Jackson and Josephine Counties, Oreg.: Mineral Resources of Oregon, vol. 1, No. 5, Oregon Bur. Mines and Geology, 1914.

GEOGRAPHY OF THE KLAMATH REGION

RELIEF AND DRAINAGE

The portions of southwestern Oregon covered in this report lie entirely within the Klamath Mountains, as defined by Diller[1] and later described by Diller,[2] Anderson,[3] and others. The region is for the most part extremely rugged and is characterized by a strong relief, which in places exceeds 3,500 feet within a distance of 2 miles. Altitudes range from over 5,000 feet on the higher peaks to sea level along the coast. The Klamath Mountains, which lie partly in California and partly in Oregon, extend north and south for over 200 miles and average about 80 miles in width. They lie west of the Cascade Range and between the coast ranges of Oregon and those of California. (See pl. 9.) The north and south boundaries are not well defined but are dependent largely upon the greater age and more complex geologic history of the Klamath Mountains; the eastern boundary is defined principally by the edges of the lavas of the Cascade Range.

The Klamath Mountains of Oregon are drained, for the most part, by the Rogue and Chetco Rivers and their tributaries. The Rogue River rises west of Crater Lake, near the crest of the Cascade Range, and flows westward by a circuitous course to the Pacific Ocean. It is joined by Bear Creek, the Applegate and Illinois Rivers, and numerous smaller tributaries. The Chetco River rises about 10 miles west of Kerby, close to the boundary line between Josephine and Curry Counties, and flows westward to the ocean through steep and rugged canyons.

---

[1] Diller, J. S., Tertiary revolution in the topography of the Pacific coast: U. S. Geol. Survey Fourteenth Ann. Rept., pt. 2, p. 408, 1894.

[2] Diller, J. S., Topographic development of the Klamath Mountains: U. S. Geol. Survey Bull. 196, pp. 9–66, 1902.

[3] Anderson, F. M., The physiographic features of the Klamath Mountains: Jour. Geology, vol. 10, pp. 144–159, 1902.

TOPOGRAPHIC FEATURES OF THE KLAMATH MOUNTAINS

The Klamath Mountains include several ridges and groups of ridges trending in various directions and with but little semblance to orderly arrangement. However, when viewed from the higher summits they present a strikingly uniform outline against the sky and, like so many other mountains that appear extremely rugged from the valleys, resemble a dissected plateau. The high summit areas, which in many places truncate upturned rocks of several ages, are remnants of an old erosion surface that Diller [4] has termed the Klamath peneplain. In general, the old surface rises gradually from north to south and from west to east. The altitude is about 2,000 feet near the mouth of the Rogue River but increases to over 4,000 feet 25 miles inland. The irregular groups of subranges and peaks that have received individual names—for example, the Siskiyou Mountains—are primarily residual ridges which divided the drainage of the principal streams during the development of the peneplain. Anderson [5] has expressed the belief that the transverse ridges are the result of cross folding after the drainage of the region had become established but, although faulting and folding have clearly played an important part in the development of the Klamath Mountains, sufficient evidence has not yet been presented to prove that most of the subranges within the mountains are other than groups of residual ridges formed by the normal erosion of an uplifted land mass. Faulting has probably accentuated the relief above the old surface in some places, but in the areas of northwestern California and southwestern Oregon that have been studied the old rocks maintain a remarkably persistent north-northeast strike, and almost everywhere the old peneplain abuts against mountains, some of which rise several thousand feet above it.

AGE AND DEVELOPMENT OF THE KLAMATH MOUNTAINS

Diller [6] has pointed out that the key to the age of the Klamath peneplain is to be found in the Neocene sediments derived from the Klamath region and laid down along the seaward border during the process of peneplanation. These sediments are well exposed in a number of places along the Oregon and northern California coasts and have been determined by Dall as belonging to the Empire formation (later Miocene). More recent correlations of the Empire formation tend to reduce the age of the peneplain and consequently

[4] Diller, J. S., Topographic development of the Klamath Mountains: U. S. Geol. Survey Bull. 196, p. 15, 1902.

[5] Anderson, F. M., Physiography of the Klamath Mountains: Jour. Geology, vol. 10. p. 150, 1902.

[6] Diller, J. S., op. cit. (Bull. 196), pp. 30-41.

the age of the Klamath Mountains and hence affect the age assignment of the older auriferous gravel in southwestern Oregon, for that is dependent, in a large measure, upon the age assigned to the Klamath peneplain.

In 1922 Howe [7] correlated the Empire formation with the standard section as worked out in California and stated that the Empire formation of Oregon is clearly of lower Pliocene age. On this basis, then, the age of the Klamath peneplain is more properly assigned to the early Pliocene, and hence the uplift of the present Klamath Mountains began during or after the middle Pliocene.

The uplift was not accomplished in one cycle but in stages interrupted by periods of relative stability. The first major uplift resulted in the cutting of broad older valleys, which in some localities are now perched as much as 2,000 feet above the present streams. A later great uplift caused younger valleys to be cut, in many places, within the older valleys. In a few places the old drainage was partly diverted by the later movements, so that some of the old channels are found at a distance from the present stream valleys. Erosion has removed the ancient gravel of most of the old channels, but in a few places—for example, near Takilma and above the Klamath River in California [8]—the gravel still remains and locally contains enough gold to be minable. Terraced forms along the present stream valleys show that the region has been unstable even during relatively recent geologic time, whereas the narrow V-shaped canyons of the present rivers indicate that the last major period of uplift may still be in progress.

Glacial lakes, oversteepened rock walls, moraines at higher altitudes, and gravel deposited by swift glacial streams at lower altitudes show that valley glaciers were developed in the higher mountains at some time after the cutting of the narrow valleys, although at no place does the glacial erosion dominate the topography. The slightly weathered condition of the boulders and the small amount of erosion that has taken place in these regions since the ice melted indicate that the glaciation in the Klamath Mountains probably occurred late in the Pleistocene epoch.

CLIMATE AND VEGETATION

Southwestern Oregon has a mild, pleasant climate characterized by warm, dry summers and cooler winters with considerable rainfall. Light snowfalls in the valleys soon disappear; heavier snows in the

[7] Howe, H. V., Faunal and stratigraphic relationships of the Empire formation, Coos Bay, Oreg.: California Univ. Dept. Geol. Sci. Bull., vol. 14, No. 3, pp. 85–114, 1922.

[8] Diller, J. S., op. cit. (Bull. 196), p. 50.

mountains lie for longer periods. The rainfall varies considerably in different parts of the region, even at points with similar altitudes. The following climatic data from three stations within a radius of 40 miles, taken from the publications of the United States Weather Bureau, give an idea of the general climatic conditions in the principal valleys.

Average precipitation, in inches, at Medford, Grants Pass, and Waldo, Oregon, 1920–30

[Altitude: Medford, 1,398 feet; Grants Pass, 956 feet; Waldo, 1,583 feet]

Station	Jan.	Feb.	Mar.	Apr.	May	June	July	Aug.	Sept.	Oct.	Nov.	Dec.	Annual
Medford	1.67	1.74	1.30	0.99	0.84	0.80	0.14	0.30	0.76	1.67	2.72	3.04	15.97
Grants Pass	3.48	4.08	2.12	1.82	.86	.69	.01	.12	.88	2.95	3.91	4.82	25.74
Waldo	a6.55	a5.84	a4.93	a3.52	a2.07	.83	.02	.20	1.38	a5.78	7.60	9.54	48.26

a Data not complete.

Average temperature, in degrees Fahrenheit, at Medford, Grants Pass, and Waldo, Oregon, 1920–30

Station	Jan.	Feb.	Mar.	Apr.	May	June	July	Aug.	Sept.	Oct.	Nov.	Dec.	Mean annual
Medford	38.0	42.3	46.3	51.2	58.7	65.9	72.3	70.5	62.7	53.5	44.9	37.6	53.66
Grants Pass	40.6	45.1	49.2	52.3	59.7	66.1	71.6	70.7	63.2	54.5	45.8	39.9	54.90
Waldo	a36.3	a40.6	a44.2	a48.1	a54.9	a62.5	68.2	67.6	60.5	a50.7	a43.5	a37.8	51.24

a Data not complete.

With the exception of the alluvium-filled valleys, most of southwestern Oregon is covered by dense growths of timber, extending from the mountains to the edges of the valley floors. By far the larger part of the trees are conifers. Douglas fir, cedar, and hemlock are the most abundant; sugar pine, yellow pine, silver fir, red fir, and spruce are more sparsely distributed. Oak, ash, maple, mountain mahogany, aspen, cottonwood, and balsam are the more abundant broad-leaved trees. The undergrowth on the timbered slopes is dense, in many places almost impenetrable and consists principally of buck brush, chinquapin, dogwood, willow, alder, laurel, yew, madrona, vine maple, soft maple, salal, huckleberry, rhododendron, manzanita, chaparral, ferns, and young conifers. The thickness of the growth depends considerably on the underlying rocks. Areas underlain by serpentine, for example, support only a scanty growth of timber but are usually covered with scattered bunches of manzanita and chaparral. Grass and plants of many varieties grow in profusion, and, where cultivated, the land is exceptionally productive.

LOCATION AND TOPOGRAPHY OF THE
TAKILMA-WALDO DISTRICT

The Takilma-Waldo district, as mapped, includes over 33 square miles in Josephine County just north of the California boundary. Takilma is the principal town, though Waldo, now deserted, was once a thriving mining town. Waters Creek, on the California & Oregon Coast Railroad 27 miles to the northeast, is the nearest shipping point. Splendid highways connect Takilma with Grants Pass, 40 miles to the northeast, and with Crescent City, a California seaport 45 miles to the southwest.

The relief near Takilma is moderate. (See pl. 10.) The ridge tops to the southwest rise about 1,400 feet above the valley floor, and Mount Hope, 2 miles southeast of Takilma, rises 2,750 feet above the valley. The district is drained by the East and West Forks of the Illinois River. The East Fork rises about 6 miles south of the California boundary and the West Fork near the line. At first both flow through narrow canyons, but about 4 miles north of the State line they come into broad flat valleys, and about 11 miles north of the line they join to form the Illinois River. This stream continues to flow northward through a wide valley for an additional 5 miles and then enters another narrow, rugged gorge through which it flows the remainder of its course to the Rogue River.

GEOLOGY

DISTRIBUTION OF ROCKS

Rocks of igneous and sedimentary origin are about equally abundant in the Takilma-Waldo district. (See pl. 11.) Sedimentary rocks prevail in the northwestern part of the mapped area, whereas igneous rocks are more abundant in the southeastern part. In addition to consolidated sediments of Paleozoic, Jurassic, Cretaceous, and late Tertiary ages, there are deposits of unconsolidated gravel which have been laid down both before and after the advent of mountain glaciation. The igneous rocks include a thick series of greenstones of probable Paleozoic age and later intrusive peridotite, now for the most part altered to serpentine. Granodiorite and other related igneous rocks are found in adjacent areas.

SEDIMENTARY ROCKS

PALEOZOIC ROCKS

Paleozoic rocks of sedimentary origin crop out in a belt about a mile wide which extends from the north end of the Takilma area to a point within half a mile of the California boundary. This belt is not composed entirely of sedimentary strata but contains a small proportion of interbedded igneous rocks. The sediments have undergone considerable metamorphism and to-day consist principally of chert, argillite, quartzite, fine-grained quartzitic conglomerate, and lentils of limestone. The chert contains abundant remains of microscopic radiolarians, which, as pointed out by Diller,[9] afford proof of the oceanic origin of the material. Limestone lentils are numerous in the Paleozoic formations of southwestern Oregon and, because of their economic value, have in the past been described in considerable detail. Diller called attention to four belts in which limestone lenses occur. These belts extend for many miles in southwestern Oregon and northwestern California, although none of the individual lenses are over one third of a mile in length, and few are over 200 feet in width. One lens crops out near Takilma in the NW. ¼ sec. 25, T. 40 S., R. 8 W. It belongs to Winchell's " west belt ",[10] which, upon fossil evidence, he considered to be of probable Carboniferous age.

Where fresh, the Paleozoic conglomerates are dark grayish green, but at the surface they are commonly bleached to a light brown. They are prevailingly fine grained, although in some of the beds the pebbles exceed an inch in length. The conglomerates grade into and are interbedded with even-grained and coarse pebbly quartzites. Rock fragments and quartz pebbles constitute the bulk of the coarser grains in the quartzites, and much of the material could properly be classified as graywacke. All the conglomerates and quartzites are well indurated, and in many places flattened quartz grains and much chlorite along slickensided surfaces reflect the intense metamorphism to which the rocks have been subjected.

The cherts are dense, light or dark gray rocks which tend to break into tabular blocks along joint surfaces. In the field they appear to be composed almost entirely of dense noncrystalline silica, but the microscope shows that they contain a large percentage of radiolarians and some quartz veinlets. Interbedded with the conglomerates and cherts are some layers of argillite. They are fine grained, have a tabular cleavage, and when breathed upon give off a strong clay odor. Because they tend to disintegrate readily, the argillite outcrops are less conspicuous than those of conglomerate and chert.

JURASSIC ROCKS

GALICE FORMATION

Rocks called by Diller a portion of the Galice formation crop out in secs. 4 and 16, T. 41 S., R. 8 W., and along the creek bottoms in secs. 29 and 31, T. 40 S., R. 8 W. Fossils found near the Almeda

[9] Diller, J. S., Mineral resources of southwestern Oregon: U. S. Geol. Survey Bull. 546, p. 15, 1914.

[10] Winchell, A. M., Petrology and mineral resources of Jackson and Josephine Counties, Oreg.: Mineral Resources of Oregon, vol. 1, No. 5, p. 36, Oregon Bur. Mines and Geology, 1914.

mine, near Galice, and on Cow Creek show that this formation occurs at about the same horizon as the Mariposa slate of the Mother Lode region in California.[11] The rocks of this formation as exposed in the Takilma-Waldo area include fine-grained conglomerate, argillite, slate, and sandstone. In other localities Diller reports variously colored chert, and mica and hornblende schists.

The sandstones are prevailingly light gray and in most places thin-bedded. Both fine-grained conglomerate and pebbly sandstone are abundant, and where the beds are steeply inclined the larger fragments are well oriented and in places flattened by pressure and adjustments that affected the rocks during deformation. Quartz grains predominate, although various kinds of rock fragments are present, including quartzite, argillite, slate, and some intensely altered fragments that may originally have been greenstone. Calcite and fine-grained quartz are abundant.

The argillites and slates are differentiated on the basis of cleavage. The slates split into thin laminae along planes independent of the original bedding; the argillites are more massive and split along bedding planes. Both are dark grayish green or black and fine-grained. Carbonaceous material is abundant in some of the shales, and in a few places poorly preserved fossil remains are found.

CRETACEOUS ROCKS
HORSETOWN(?) FORMATION

Rocks of Cretaceous age, classed by Diller as Horsetown, crop out in the northwestern part of the Takilma-Waldo area. The fossil evidence does not definitely place them in either the upper Horsetown or lower Chico, but for convenience they are described under the name originally given to them. The Cretaceous rocks near Takilma are remnants of a once widespread formation that Diller [12] states may have formed a continuous blanket on the older rocks over almost the whole of the Klamath Mountains. Winchell [13] feels that the evidence for such a widespread distribution is insufficient, and Anderson [14] holds that there was a considerable land mass in this region during late Cretaceous time and that no connection existed between the Sacramento Valley and the Cretaceous basin of southern Oregon during the deposition of the Chico rocks. Whether or not the California and Oregon basins were connected, the available information indicates that at least parts of the present Klamath Mountains were probably not submerged during the late Cretaceous.

[11] Diller, J. S., op. cit. (Bull. 546), p. 17.
[12] Idem, p. 18.
[13] Winchell, A. M., op. cit., p. 34.
[14] Anderson, F. M., The physiographic features of the Klamath Mountains: Jour. Geology, vol. 10, p. 152, 1902.

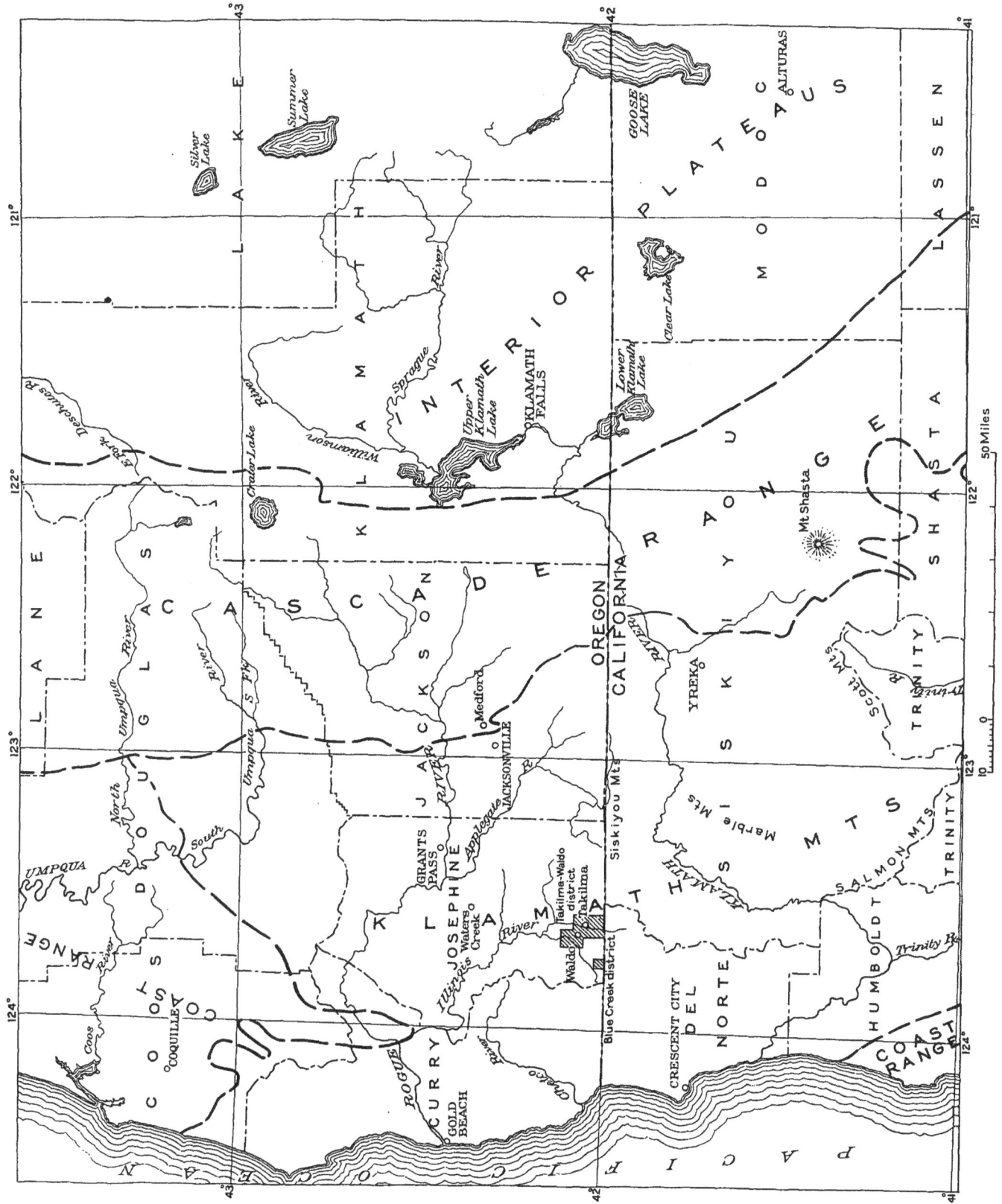

INDEX MAP OF SOUTHWESTERN OREGON AND NORTHWESTERN CALIFORNIA, SHOWING PHYSIOGRAPHIC DIVISIONS.

U. S. GEOLOGICAL SURVEY

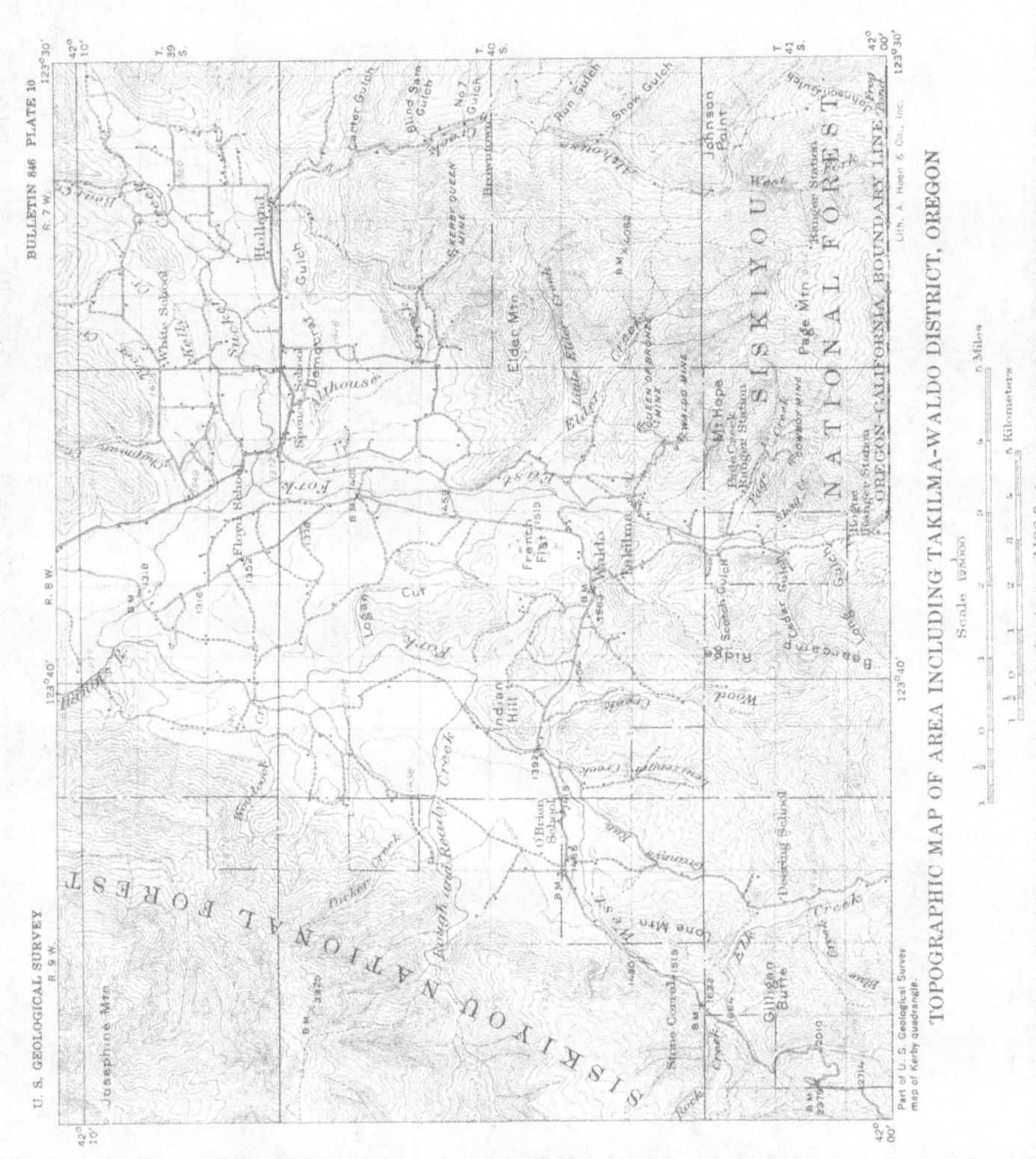

Part of U. S. Geological Survey
map of Kerby quadrangle.

Scale 125000

5 Miles

5 Kilometers

TOPOGRAPHIC MAP OF AREA INCLUDING TAKILMA-WALDO DISTRICT, OREGON

Lith. A. Hoen & Co., Inc.

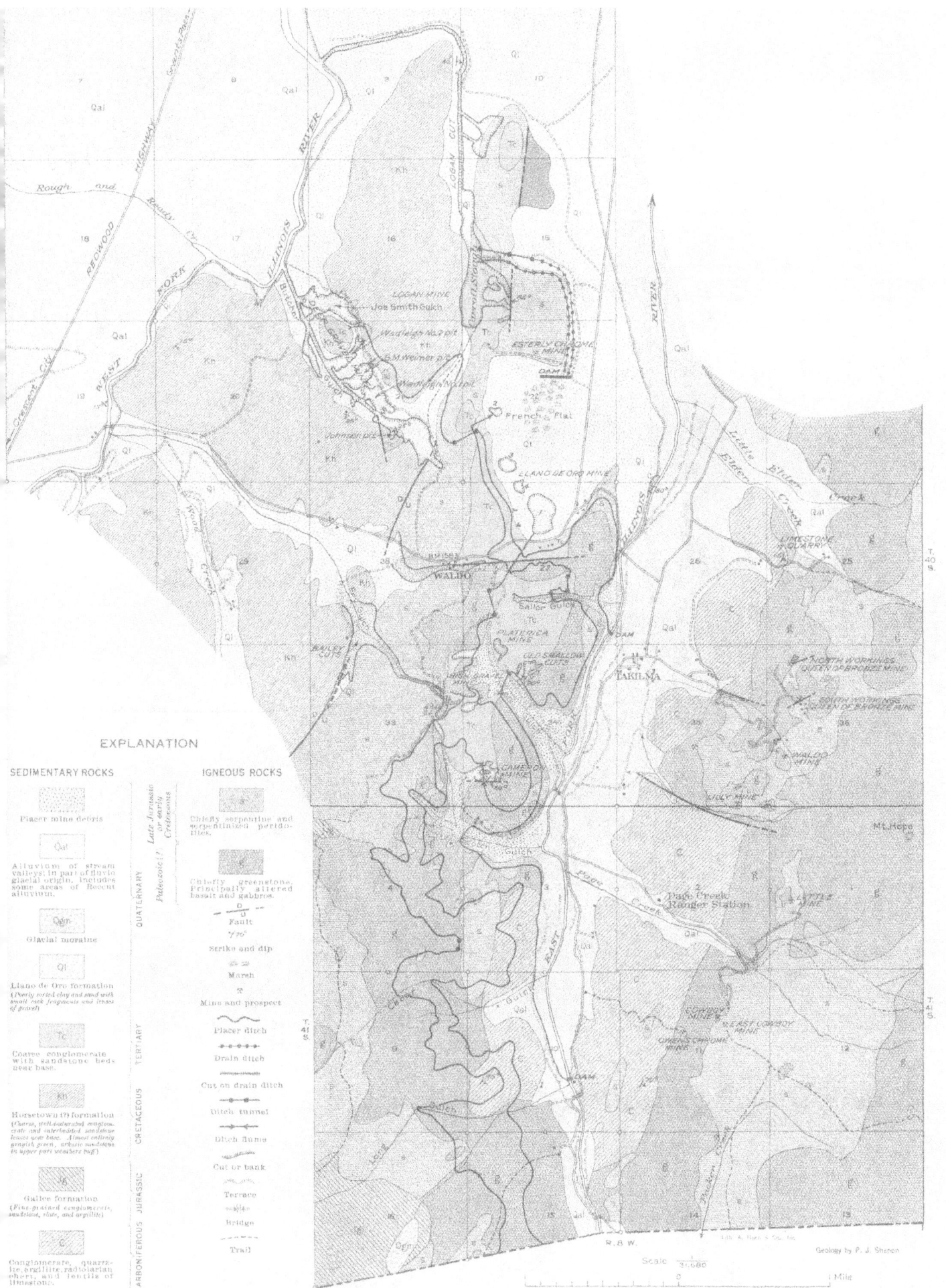

EXPLANATION

SEDIMENTARY ROCKS

Placer mine debris

Qal

Alluvium of stream
valleys; in part of fluvio-
glacial origin. Includes
some areas of Recent
alluvium.

Qgm

Glacial moraine

Ql

Llano de Oro formation
(Poorly sorted clay and sand with
small rock fragments and lenses
of gravel)

Tc

Coarse conglomerate
with sandstone beds
near base.

Kh

Horsetown (?) formation
(Coarse, well-indurated conglomerate and subordinate sandstone
lenses near base. Almost entirely
greenish green; arkosic sandstone
in upper part weathers buff)

G

Gallice formation
(Fine-grained conglomerate,
sandstone, slate, and argillite)

C

Conglomerate, quartz-
ite, argillite, radiolarian
chert, and lentils of
limestone.

IGNEOUS ROCKS

Chiefly serpentine and
serpentinized peridotite.

Chiefly greenstone.
Principally altered
basalt and gabbros.

Fault

Strike and dip

Marsh

Mine and prospect

Placer ditch

Drain ditch

Cut on drain ditch

Ditch tunnel

Ditch flume

Cut or bank

Terrace

Bridge

Trail

QUATERNARY
TERTIARY
CRETACEOUS
JURASSIC
CARBONIFEROUS

Late Jurassic
or early
Cretaceous
Paleozoic(?)

Scale 1:31,680

0   1 Mile

Geology by P. J. Shenon

GEOLOGIC MAP OF THE TAKILMA-WALDO DISTRICT, JOSEPHINE COUNTY, OREGON

A. CRETACEOUS CONGLOMERATE AND SANDSTONE LENSES 700 FEET NORTH OF
O'BRIEN BRIDGE, IN SW. ¼ SEC. 19, T. 40 S., R. 8 W.

B. SPHEROIDAL WEATHERING IN CRETACEOUS SANDSTONE ALONG HIGHWAY IN
NW. ¼ NE. ¼ SEC. 29, T. 40 S., R. 8 W.

A. TERTIARY CONGLOMERATE RESTING ON GREENSTONE AT HIGH GRAVEL MINE.

Photograph by J. H. Maxson.

B. TERTIARY CONGLOMERATE WITH SANDSTONE LENSES AT PLATERICA MINE.

Photograph by J. T. Pardee.

A. GRAVEL OF LLANO DE ORO FORMATION RESTING ON TERTIARY CONGLOMERATE
AT LLANO DE ORO MINE.

Photograph by G. M. Esterly.

B. SERPENTINE BEDROCK AT LLANO DE ORO MINE.

Photograph by G. M. Esterly.

The lower beds of the Cretaceous rocks exposed near Takilma are largely coarse conglomerates with some interbedded sandstones; the upper beds are almost totally sandstones. (See pl. 12.) The conglomerates grade upward into massive sandstone beds, which in places are fossiliferous. The conglomerates are composed principally of boulders and pebbles of dark-gray quartzitic sandstone and lesser amounts of slate, greenstone, and granitic rocks. Some of the boulders have diameters exceeding a foot but average less than 6 inches. Conglomerate beds somewhat resembling in appearance those of the gold-bearing Tertiary rocks are exposed along the West Fork of the Illinois River about 1,500 feet east of the mouth of Rough and Ready Creek. No information was obtained regarding their gold content, although some small cuts in the SW. ¼ sec. 17, T. 40 S., R. 8 W., may represent an attempt to wash them. Sandstone makes up the bulk of the Horsetown (?) formation. It is massive, is grayish green where fresh, and during weathering has a strong tendency to develop spheroidal structure (pl. 12, *B*). Concretions filled with loose brown sandy material are numerous. The sandstone is remarkably even-grained over considerable areas, and in most places bedding is not readily discernible. Angular quartz and some feldspar grains, for the most part less than 0.10 millimeter long, constitute about 30 per cent of the prevailing sandstone, and highly altered material, in part chloritized rock fragments, makes up the remainder. The following fossils from the sandstone were identified by T. W. Stanton:

15411. Lot 1. Sandstone, Logan cut, near quarter corner between secs. 9 and 10, T. 40 S., R. 8 W.:

    Pecten operculiformis Gabb.
    Pteria sp.
    Meekia sp.
    Pleuromya papyracea Gabb.
    Solecurtus? sp.
    Undetermined pelecypods.
    Lunatia sp.

15412. Lot 2. Sandstone, road cut near quarter corner between secs. 20 and 29, T. 40 S., R. 8 W.:

    Pecten operculiformis Gabb.
    Inoceramus sp., fragments.
    Leda sp.
    Meekia? sp.
    Pleuromya papyracea Gabb.
    Corbula sp.
    Undetermined pelecypods.

According to Mr. Stanton,

These fossils belong to a Cretaceous fauna whose stratigraphic classification has not been finally decided, though it is either upper Horsetown or lower Chico. The particular area from which these collections came was mapped by Diller as Horsetown in United States Geological Survey Bulletin 546, 1914, on pale-ontologic evidence practically identical with that contained in the above col-lections. The fact must be admitted, however, that even in the Sacramento Valley, which is the type area of both the Horsetown and the Chico, the bound-ary between these two formations is not very definite and as applied has prob-ably varied at different localities. When finally adjusted it will probably be found that the fauna represented by these collections belongs only below or above the boundary and not on both sides of it. Meanwhile it makes little difference whether this sandstone in southwestern Oregon is called Horsetown or Chico.

TERTIARY ROCKS

CONGLOMERATE

Several bodies of a gold-bearing conglomerate occur in the north-ern part of the Takilma-Waldo district, chiefly on the divide between the East and West Forks of the Illinois River, from the Cameron mine northward for a distance of about 4 miles. Evidently these bodies are the erosion remnants of a once continuous formation that had a maximum thickness of at least 400 feet between Allen and Sailor Gulches.

The conglomerate is made up of well-rounded cobbles and boulders in a matrix of sandy clay. A few of the boulders are as much as 3 feet in diameter, but most are less than 1 foot. Weathering has decomposed the formation to such an extent that most of the cobbles fall to pieces when released from the mass.

The cobbles and nuclei of boulders that remain firm consist mainly of greenstones, but a few are of chert or other fine-grained siliceous sediments. The matrix is abundant, and in places, particularly the lower part of the formation, there are lenslike bodies of sandstone. (See pl. 13, B.)

The conglomerate rests on a surface that in different places is eroded across greenstone (pl. 13, A), serpentine, and Cretaceous sandstone. Near the Cameron and Osgood mines this surface is about 600 feet higher than the valleys adjacent to it. North-ward the conglomerate occupies successively lower positions, until at the Llano de Oro mine part of it is lower than the present East Fork Valley. Still farther north on a hill southeast of the Logan cut, a patch is 100 feet or more above the adjacent valley.

Deformation of the Tertiary conglomerate by faulting is clearly shown at the Cameron placer cut, where the formation is cut off on the south by an east-west fault that dips 65° N. The fault plane, accompanied by gouge and slickensides, is well exposed, and the conglomerate is indicated to be downthrown on the north at least 200 feet. Exposures at the Osgood and Platerica mines show several small faults that cut both the conglomerate and the bedrock. These fractures pass through the boulders and cobbles instead of around them, a fact that indicates the matrix to have been firmly cemented when the deformation occurred. In sec. 21, T. 40 S., R. 8 W., on the south side of the Deep Gravel mine, a pit worked by Charles Johnson in 1925 exposes a small isolated body of the Tertiary conglomerate resting unconformably on Cretaceous sandstone. The conglomerate at this place dips at a moderate angle westward. The successively lower positions of the bodies of conglomerate on the north side of Allen Gulch and south of the Llano de Oro mine are interpreted as most probably due to faulting. (See fig. 22.)

Diller [15] believed the conglomerate under consideration to be a marine deposit of Cretaceous age. His interpretation was based on the apparent relation of the rocks exposed in the Logan cut, near the quarter corner between secs. 15 and 16, T. 40 S., R. 8 W. At that locality (pl. 11), the conglomerate is exposed at the east side of

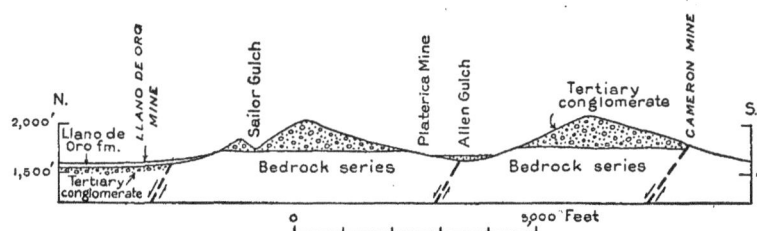

FIGURE 22.—Section illustrating relation of Tertiary conglomerate to bedrock, Takilma-Waldo district, Oreg.

Carroll Slough and lies on a serpentine bedrock. West of the slough is a low ridge of Cretaceous sandstone. If the conglomerate persists in the direction of its dip it must pass beneath the sandstone, and if their relations are conformable the conglomerate is of course the older. At present (1930) the contact of the two formations is concealed by an accumulation of tailings and a natural deposit of alluvium along the west side of the slough. Whether the contact was actually in view at the time of Diller's examination or concealed, as at present, is not known. In the light of additional facts observed in the more recent examination, it is concluded that despite the apparent relation of the formations at the Logan mine, the conglomerate is a river deposit resting unconformably on the Cretaceous sandstone. The evidence on which this interpretation is based consists of (1) lack in the conglomerate of the assortment characteristic of beach deposits; (2) a linear arrangement of the conglomerate bodies (pl. 11) that suggests a main stream flowing north or south with a west branch joining it at Butcher Gulch (Deep Gravel mine); (3) remnants of the conglomerate resting on the

[15] Diller, J. S., Mineral resources of southwestern Oregon; U. S. Geol. Survey Bull. 546, pp. 93–95, 1914.

Cretaceous sandstone at places near the Deep Gravel mine, particularly in the Johnson pit of 1925, where the unconformable relation of the conglomerate and sandstone is actually in view. It should be noted that the exposure last mentioned and some others near the Deep Gravel mine were not to be seen when Diller last visited the region.

The conglomerate, which is unconformably overlain by Pleistocene alluvium, herein named the Llano de Oro formation, is moderately deformed and deeply weathered and is therefore considered to be as old as Tertiary. A tentative correlation is suggested between the Tertiary conglomerate of the Takilma-Waldo district and the Miocene gold-bearing gravel of Trinity River, California, described by Diller [16] as river deposits formed in old valleys eroded deeply in the Klamath peneplain. As described, those deposits are not unlike the Tertiary conglomerate in texture, cementation, and weathering.

QUATERNARY ROCKS
LLANO DE ORO FORMATION

At the Llano de Oro mine the Tertiary conglomerate passes unconformably beneath a deposit of alluvium (pl. 11) that extends northward beyond the limit of the district. The deposit here named the Llano de Oro formation makes a low terrace bordering the East Fork of the Illinois River and at the extreme north of the district occupies the entire space between that stream and the West Fork. Thence it extends up the West Fork as a bordering terrace to Butcher Gulch and beyond. In addition, it occupies the small valley of Butcher Gulch up to the head of the Deep Gravel mine and floors two or more tributary valleys farther south, including the one beyond Indian Hill that leads up to Waldo. Small bodies of alluvium that have been mined in Sailor, Allen, and Scotch Gulches are thought to be included with the same deposit.

The Llano de Oro formation consists mainly of poorly assorted clay and sand with small rock fragments and lenses of gravel. It contrasts strongly with the later Pleistocene gravel described below. It is of the type produced by slow and uninterrupted erosion and probably represents the early part of the Wisconsin stage of Pleistocene time. It is of interest to note that miners are reported to have found human implements in this gravel, and at least one of the reports seems to be fairly well authenticated. This is a granite mortar that, as described by Kemp, was found in the Deep Gravel mine at a depth of 80 feet.[17]

[16] Diller, J. S., The auriferous gravels of the Trinity River Basin, Calif.: U. S. Geol. Survey Bull. 470, pp. 11–29, 1910.

[17] Kemp, J. F., Recent interesting discovery of human implements in an abandoned river channel in southern Oregon [abstract]: New York Acad. Sci. Annals, vol. 17, pp. 606–608, 1907.

ALLUVIAL FAN OF ROUGH AND READY CREEK

A coarse bouldery deposit along Rough and Ready Creek forms a large alluvial fan which has crowded the West Fork of the Illinois River into the Llano de Oro formation on its right bank, thus forming a cliff-like terrace from 30 to 50 feet high. The boulders and cobbles are composed chiefly of peridotite and other basic rocks, as shown in a cut on the Redwood Highway about a mile south of the creek crossing. The material is weathered to a deep-red clay soil, and the coarser-textured rocks are surrounded by completely softened shells from 1 inch to several inches thick. This exposure is on a small ridge that apparently marks the original top of the alluvial fill. Adjacent areas have been lowered 20 or 30 feet by stream erosion that occurred mostly after the rocks had been weathered as described, and in these areas the material appears relatively fresh.

The size of the deposit and the number of very large boulders in the gravel indicate that the material was deposited by a stream much larger than the present one. The sequence of events indicated—namely, rapid deposition of alluvium followed by a moderate degree of weathering and later erosion by a clarified stream—suggests that the original deposit was transported by a stream derived from melting glaciers. This idea is supported by the fact that a cirquelike basin containing a small lake is shown by the United States Geological Survey topographic map of the Kirby quadrangle at the head of one of the forks of Rough and Ready Creek. It is concluded that the deposit is of late Pleistocene age, but the degree of weathering indicates it to be probably somewhat older than the latest Wisconsin.

GRAVEL DEPOSITS OF THE EAST FORK OF THE ILLINOIS RIVER

Rather coarse alluvium, correlated here with the deposit on Rough and Ready Creek, fills the valley of the East Fork of the Illinois River and some of its tributaries in the vicinity of Takilma. In it are well-developed terraces, usually less than 10 feet high, but two in particular are fairly continuous. This deposit, like the alluvial fan of Rough and Ready Creek, fills a valley in part eroded in the Llano de Oro formation. The boulders are principally greenstones, serpentine, and granitic rocks, with lesser amounts of quartzite, argillite, and chert. Like the deposit on Rough and Ready Creek, this gravel is believed to have been transported, in part at least, by streams derived from melting glaciers during late Pleistocene time. At the mouths of the upper tributary valleys the boulders are very large and the material poorly sorted. Furthermore, cirques with lakes occur at the headwaters of the East Fork, and a small moraine deposited by a glacier in the same drainage basin is found near the south border of the district.

IGNEOUS ROCKS

Igneous rocks are most abundant in the southeastern part of the Takilma-Waldo district. They include greenstones of several varieties and serpentinized peridotite (or serpentine). Granodiorite float was found in sec. 1, T. 41 S., R. 8 W., although no outcrops were seen. The greenstones include dense porphyritic and fragmental metabasalts, medium-grained metagabbro, porphyritic metagabbro, and probably meta-andesite and metadiorite. The serpentine and greenstones were mapped separately, but because of the thick soil in most places no attempt was made to differentiate the varieties of greenstone.

GREENSTONES

The term "greenstone" has been applied to fine and medium grained altered igneous rocks which, owing to the development of considerable chlorite and epidote, are characterized by a green color. These rocks are of particular interest in southwestern Oregon, because many of the economically important ore deposits are enclosed in them. Two varieties prevail in the Takilma-Waldo district—a very fine grained altered basalt (or metabasalt) which in most places partly retains a well-developed basaltic texture, and a medium-grained altered gabbro (or metagabbro) of similar composition. Porphyritic metabasalt and altered fragmental rocks of basaltic composition are less abundant.

The greenstones have undergone considerable metamorphism, which has resulted in the formation of a new mineral assemblage. Analyses show the rocks to be considerably more calcic than the composition of the feldspars would indicate, and in some of them there is good textural evidence of recrystallization. No feldspars were found to be more calcic than Ab_{70}, and yet the analyses indicate a basaltic composition. Diller [18] has also noted the lack of agreement between the analyses of the greenstones and the composition of their feldspars in the vicinity of Riddle, Oreg., and several writers have discussed in detail the effect of dynamothermal metamorphism upon the mineral composition of rocks.

The age of the greenstones in the Takilma-Waldo district can not be definitely fixed. Part of them at least are interbedded with the Paleozoic sedimentary rocks, and, as pointed out by Diller,[19] others are of Mesozoic age. All are intruded by serpentine which is believed to be of pre-Chico age.

---

[18] Diller, J. S., U. S. Geol. Survey Geol. Atlas, Riddle folio (No. 218), p. 5, 1924.

[19] Diller, J. S., Mineral resources of southwestern Oregon: U. S. Geol. Survey Bull. 546, p. 19, 1914.

Metabasalts are the most abundant greenstones in the Takilma-Waldo district. Their flow structure shows that some are interbedded with sedimentary rocks and no doubt some are intrusive into the sedimentary rocks, but by far the greater part occur as a separate unit made up of flows and fragmental material. Grayish green is the prevailing color, but in some places abundant epidote and quartz impart a mottled pistachio-green color to the rocks. A light-gray rock, occurring in secs. 2 and 13, T. 41 S., R. 8 W., and resembling fine-grained quartzite is proved by the microscope to be of igneous origin and is hence mapped with the metabasalt.

Most of the metabasalts have a basaltic texture, although some are porphyritic. Lath-shaped feldspars, with average cross sections of about 0.5 by 0.8 millimeter, constitute about 60 per cent of the fresher rocks. The mineral composition of the remaining 40 per cent varies in different places but includes augite, diopside, clinoenstatite, ilmenite, magnetite, pyrite, chlorite, epidote, uralite, leucoxene, saussurite, clay minerals, and limonite. Sericite is present in small amounts. Hornblende is abundant in some of the porphyritic greenstones.

The feldspars have undergone hydrothermal alteration but are relatively much fresher than the ferromagnesian minerals. Sodic plagioclase is the most abundant feldspar. It varies somewhat in composition in the different metabasalts of the Takilma-Waldo district but is not known to be more calcic than Ab_{70} and averages about Ab_{85}. Orthoclase accounts for less than 5 per cent of the rock. Little or no primary quartz is present, but some quartz formed by the breaking down of other minerals is generally evident, and in some places considerable quartz, as well as calcite and pyrite, has been introduced. Green fibrous chlorite, epidote, and uralite are the most abundant dynamothermal minerals. Some clay minerals and limonite have been formed by processes of surface weathering.

Near the ore deposits the metabasalts are changed to a mass composed chiefly of green chlorite, fine-grained quartz, calcite, and sulphides. The following analysis is representative of the prevailing metabasalts of the district. It corresponds closely with an analysis of a rock from the " north end adit, Queen of Bronze mine " which Winchell [20] has termed auganite.

[20] Winchell, A. N., Petrology and mineral resources of Jackson and Josephine Counties, Oreg.: Mineral Resources of Oregon, vol. 1, No. 5, p. 51, Oregon Bur. Mines and Geology, 1914.

Composition of metabasalt from the SW. ¼ SW. ¼ sec. 36, T. 40 S., R. 8 W., Takilma, Oreg.

[J. G. Fairchild, analyst]

Analysis		Norm (III.5.3.4.)		Mode (estimated)	
SiO_2	49. 31	Quartz	0. 00	Quartz	0
Al_2O_3	13. 72	Orthoclase	2. 78	Oligoclase (Ab_{85})	55
Fe_2O_3	2. 84	Albite	37. 20	Orthoclase	2
FeO	10. 64	Anorthite	16. 12	Diopside	12
MgO	5. 50	Diopside	11. 90	Magnetite	3
CaO	6. 30	Hypersthene	13. 44	Leucoxene	6
Na_2O	4. 44	Olivine	5. 17	Pyrite	1
K_2O	. 53	Magnetite	4. 18	Fine-grained, scaly, limonite-stained dynamothermal alteration product	21
H_2O+	3. 29	Ilmenite	4. 86		
H_2O-	. 37	Apatite	. 31		100
TiO_2	2. 57	Pyrite	. 72		
P_2O_5	. 19	H_2O+	3. 29		
S(total)	. 21	H_2O-	. 37		
MnO	. 14				
	100. 05		100. 34		

METAGABBRO

Metagabbro crops out in irregular areas and is probably intrusive into the metabasalts. It is a medium to a fine grained rock in which the light and dark minerals appear megascopically to be in about equal proportions. The microscope generally shows the less altered rock to consist principally of lath-shaped plagioclase and pyroxene minerals, and their alteration products, in proportionate amounts of about 5 to 4. Lath-shaped oligoclase, averaging about 0.7 by 0.3 millimeter in cross section, is the prevailing feldspar. The pyroxene constituents vary somewhat from place to place. Augite prevails, but clinoenstatite is in places abundant. Hornblende occurs in lesser amounts and reaches a maximum of about 10 per cent. Original quartz is sparsely distributed throughout the rock but does not exceed 5 per cent and in most places is present in proportions of less than 1 per cent. Ilmenite (partly altered to leucoxene) makes up over 5 per cent of some of the metagabbros. In places it has developed beautiful skeletal patterns. The metagabbro has everywhere undergone dynamothermal alteration. Chlorite is the most abundant recognizable alteration product, but epidote and sericite occur in smaller amounts. The ferromagnesian minerals have undergone the greatest changes, although some sericitization and alteration to a very fine grained product are usually evident in the feldspars. Like the metabasalts, the metagabbro near ore bodies is changed to a rock composed principally of chlorite, fine granular quartz, calcite, and sulphides.

The following analysis of a metagabbro from Long Gulch, in the NW. ¼ sec. 16, T. 41 S., R. 8 W., illustrates the composition of these rocks from the Takilma-Waldo district:

Composition of metagabbro from the NW. ¼ sec. 16, T. 41 S., R. 8 W., Takilma-Waldo district, Oregon.

[J. G. Fairchild, analyst]

Analysis		Norm (III.5.3.4)		Mode (estimated)	
SiO_2	49.60	Quartz	0.00	Quartz	1
Al_2O_3	14.10	Orthoclase	6.12	Orthoclase	5
Fe_2O_3	1.77	Albite	25.15	Oligoclase (Ab_{74})	35
FeO	8.40	Anorthite	21.96	Clinoenstatite	10
MgO	8.34	Diopside	6.77	Hornblende	10
CaO	7.10	Hypersthene	27.20	Magnetite	1
Na_2O	2.96	Olivine	2.04	Leucoxene	5
K_2O	1.05	Magnetite	2.55	Chlorite	10
H_2O-	.34	Ilmenite	2.28	Sericite	5
H_2O+	3.37	Apatite	.31	Calcite	2
TiO_2	1.25	Calcite	1.80	Pyrite	1
CO_2	.80	Pyrite	.12	Fine-grained, scaly dynamothermal product	15
P_2O_5	.18	H_2O+	3.37		
S	.05	H_2O-	.34		
Cr_2O_3	.05				100
MnO	.14		100.01		
	99.50				

Volcanic gas equals 1.9 milliliters per gram of rock.

SERPENTINE

Serpentine (or serpentinized peridotite) crops out in large areas in southwestern Oregon and northern California. It is abundant near Takilma, where it incloses copper and chromite deposits and is the probable source of platinum found in the placer mines. Hence it is of considerable economic interest. The outcrops are as a rule fairly well defined by sparse vegetation and a dark-red soil, although red soil is also developed by the weathering of greenstones, particularly near serpentine contacts. In general the outcrops are more rounded than those of the other consolidated rocks, but in places they are rough. The serpentine is black or very dark green and is generally characterized by slick curved faces. Serpentinization is well advanced in all the peridotite rocks near Takilma, but in the less altered rocks both enstatite and olivine occur. Crystals of bastite (altered enstatite) are usually visible in hand specimens. Antigorite, the platy variety of serpentine, is the most abundant mineral, although chrysotile, the fibrous variety, is common, particularly along fractures. Iddingsite is present in small amounts, and magnetite and chromite are disseminated through the serpentine as grains and irregular patches. In some places chromite is found in bodies large enough to be minable under favorable conditions. Lenses of diopside, some several feet long, are not uncommon.

Serpentine is usually derived from the hydrothermal alteration of peridotite or other ultrabasic igneous rocks. Benson[21] has ad-

[21] Benson, W. N., The origin of serpentine, a historical and comparative study: Am. Jour. Sci., 4th ser., vol. 46, pp. 693-729, 1918.

vanced the view that the larger antigorite and chrysotile serpentine masses are due to the alteration of pyroxene peridotites through the agency of magmatic waters belonging to the same cycle of igneous activity as the ultrabasic rocks themselves. Keep[22] deduced from field evidence the hypothesis that the serpentinization of the ultrabasic rocks of the asbestos deposits of Shabani, Southern Rhodesia, was due to the action of the magmatic waters accompanying the intrusion of the (later) granite batholith of the Lundi Native Reserve. Wells[23] treated olivine, in the laboratory, with various water solutions of Na_2CO_3, $NaHCO_3$, $NaCl$, Na_2SO_4, K_2CO_3, K_2SO_4, water glass, and CO_2 at temperatures of 100° to 600° C. and pressures of 1 to 310 atmospheres, but did not produce serpentine. He is inclined, therefore, to think that the hydrothermal alteration of olivine to serpentine is a late magmatic process taking place at temperatures exceeding 520° C.

The peridotites of the Takilma-Waldo district have been subjected, during several periods, to intense thermal and dynamic processes, but the evidence available at this time is not sufficient to warrant a statement as to which process or processes caused the serpentinization. No large bodies of intrusive igneous rocks younger than the serpentine crop out in this area, although, as such bodies crop out in adjacent areas, they are probably not far beneath the surface in the Takilma-Waldo district. Hydrothermal alteration has certainly been a factor in the formation of the serpentine, as such alteration is most complete near the ore deposits, but because the peridotite rocks have been subjected to intense deformation after their intrusion, dynamothermal processes should not be ignored as a possible factor in the serpentinization.

The age of the serpentine at Takilma is not very closely defined. It is tentatively classified as late Jurassic or early Cretaceous. The Horsetown(?) formation is believed to lie unconformably over the serpentine because at no place was serpentine observed cutting it, but no markers are known in this district by which the lower limit may be fixed.

In the Riddle region Diller[24] has assigned the serpentine to the late Jurassic or early Cretaceous, whereas in the Port Orford region serpentine cuts the Myrtle formation, of Lower Cretaceous age, and does not intersect the Arago formation, of Eocene age, and therefore, the intrusion in the Port Orford region was believed by Diller[25] to have occurred some time during the later portion of the Cretaceous.

[22] Keep, F. E., The geology of the Shabani mineral belt, Belingwe district: Southern Rhodesia Geol. Survey Bull. 12, p. 82, 1929.

[23] Wells, F. G., The hydrothermal alteration of serpentine: Am. Jour. Sci., 5th ser., vol. 18, pp. 35–52, 1929.

[24] Diller, J. S., U. S. Geol. Survey Geol. Atlas, Riddle folio (No. 218), p. 4, 1924.

[25] Diller, J. S., U. S. Geol. Survey Geol. Atlas, Port Orford folio (No. 89), p. 4, 1903.

DEFORMATION

The geologic structure of southwestern Oregon is very complex, and, because detailed information is lacking in much of the region, it has been described only in a general way. The strata older than the Cretaceous in general strike north-northeast and dip southeast at steep angles. Diller [26] has called attention to the fact that the beds increase in age to the southeast, whereas normally the reverse order should hold, and states that the apparent reversal of the natural order is due either to folding and overturning of the strata or to faulting, by which the older rocks are made actually or apparently to overlie the younger. He further states that both folding and faulting very probably have contributed to the complex structure and that the most evident line of faulting noted in the region crosses it from northeast to southwest in the vicinity of Waldo and Kerby, where the Jurassic strata appear to pass beneath the Devonian. Hershey's reconnaissance map of Del Norte County, Calif.,[27] shows a fault, the Orleans fault, extending north and south across the Klamath Mountains for a distance of about 50 miles. According to his map the fault should cross the Oregon boundary near the western edge of the Takilma-Waldo district, apparently in line with the faulting noted by Diller. Although it can not be definitely stated that the Orleans fault crosses the Takilma area, there is definite evidence that the Horsetown(?) formation is in fault contact with serpentine just west of Waldo. Normally, the lower portions of the Horsetown(?) formation in the Takilma-Waldo district are largely coarse conglomerate beds, which grade upward into sandstone. However, just west of Waldo, where the formation is in contact with serpentine, the conglomerate beds are missing and the serpentine is in direct contact with the sandstone. Deformation has affected the older rocks during several periods. The most intense deformation was associated with the mountain-building epoch at the end of the Jurassic. Later deformation followed the deposition of the Horsetown (?) formation, which in places is faulted and considerably folded, although less so than the older rocks. Still later deformation is evident in the gently deformed Tertiary conglomerate near Takilma and in folded Tertiary rocks in areas outside of the Takilma district. Hershey [28] described two post-Cretaceous periods of deformation in the Klamath Mountains of California, one in the early Tertiary and one at the beginning of the Quaternary.

---

[26] Diller, J. S., U. S. Geol. Survey Bull. 546, pp. 21–22, 1914.

[27] Hershey, O. H., Del Norte County geology: Min. and Sci. Press, vol. 102, p. 468, 1911.

[28] Hershey, O. H., Structure of the southern portion of the Klamath Mountains, Calif.: Am. Geologist, vol. 31, No. 4, p. 232, 1903.

ORE DEPOSITS

CLASSIFICATION

The ore deposits of the Takilma-Waldo district include lodes and placers. Both have been very productive in the past, and the information gathered during the present investigation indicates that both will continue to produce abundantly in the future when there is a demand for their products. During comparatively recent years the lode mines have been credited with a production of about $1,700,000, and the minimum production of the placer mines, including that of the early days, is estimated at $4,000,000. The lode deposits have produced chiefly copper. One small chromite deposit has been mined and is credited with a small production, and several other small chromite deposits are known to exist. The placer mines, which contain both gold and platinum, have in recent years been the most productive in the State.

COPPER MINES

GENERAL FEATURES

Copper deposits have been known in the Takilma district since the early sixties, but very little ore was produced until after 1900. Six productive mines and several prospects near Takilma are located within a narrow zone extending north and south for over 3 miles on the east side of the East Fork of the Illinois River. All the deposits have been found close to greenstone-serpentine contacts, and most of them in greenstone. The Cowboy deposit, which is enclosed in serpentine, is the principal exception. The copper deposits occur as irregular bodies and lenses and, except where oxidized near the surface, consist of sulphides that have been deposited in and along fractures of the enclosing rocks. The hypogene (primary) sulphides include chalcopyrite, cubanite (copper-iron sulphide), sphalerite, pyrite, pyrrhotite, and, at the Cowboy mine, cobaltite (cobalt-arsenic sulphide). The sequence of mineralization is not the same in all the deposits, but the sulphide minerals are everywhere associated with altered wall rocks, quartz, and calcite as gangue materials. Oxidation and sulphide enrichment have been important processes only close to the surface. Postmineral movements are made evident in all the mines by slickensided surfaces and displacements of ore bodies.

Because the deposits in greenstones differ in several respects from those enclosed in serpentine, they are discussed separately. The Queen of Bronze, Waldo, Lilly, and Lyttle mines are included in the first group; the Cowboy is the only important deposit in serpentine.

COPPER DEPOSITS IN GREENSTONE

QUEEN OF BRONZE MINE

LOCATION AND HISTORY

The Queen of Bronze mine is in the NW. ¼ sec. 36, T. 40 S., R. 8 W., about 1½ miles by road east of Takilma, on a prominent ridge between Elder Creek and the East Fork of the Illinois River, at an altitude of about 2,400 feet. A winding road, easily passable except during periods of heavy rains, connects the mine with Takilma.

Copper was discovered in the Takilma district [29] in 1860 by a miner named Hawes, and the Queen of Bronze mine was located October 22, 1862, by P. Androit. Little development work was done, however, until 1903, when C. L. Tutt and associates, of Colorado Springs, bought the property. Their company, known as the Takilma Smelting Co., erected a smelter of the semipyritic type in 1904. It had a capacity of 100 tons a day and operated more or less continuously until 1910. More than 20,000 tons of ore, with an average copper content of 8½ per cent, was smelted,[30] and in 1907 the mattes from the smelted ores contained about 40 per cent of copper. Lessees shipped a few thousand tons of ore during the period from 1910 to 1916. In April, 1916, the property was sold to John Twohy, John Hampshire, and others, with Hampshire acting as trustee. They operated continuously, under the direction of Roy B. Clark, until 1919, when the price of copper became too low for profitable mining. During this period of 3 years, according to G. E. Stowell, mining engineer, the mine shipped 183 cars (9,992.3 tons) of ore varying in metal content from 5.16 to 16.33 per cent of copper and from 0.04 to 0.44 ounce of gold to the ton. From 1918 to 1928 lessees made intermittent shipments. John Hampshire obtained a lease on the property in June, 1928, and operated until May 31, 1929, when the Queen of Bronze Mining Co. was incorporated. The new company continued to ship ore until May, 1930, when operations were suspended because of the low copper market. During the period from June, 1928, to May, 1930, 1,552.941 tons of ore was shipped.[31] On the basis of the above figures, the total production of the Queen of Bronze mine is estimated at about 35,000 tons of ore with a gross value of approximately $1,350,000.

---

[29] Browne, J. R., Mineral resources of the States and Territories west of the Rocky Mountains for 1866, p. 141, 1867.

[30] Diller, J. S., Mineral resources of southwestern Oregon: U. S. Geol. Survey Bull. 546, p. 83, 1914.

[31] Pohlman, Edward (secretary Queen of Bronze Mining Co.), letter of April 24, 1931.

Ore has been mined at the Queen of Bronze from what are known as the north-end workings and south-end workings. They are about 1,100 feet apart. The north-end workings (pl. 15) include over 7,000 feet of drifts and crosscuts, a large open pit, numerous stopes, and several hundred feet of raises and winzes, all of which are restricted to a roughly circular area about 500 feet across and most of which, with the exception of some openings on the upper levels, are accessible. The deepest level is about 225 feet below the open pit. The larger ore bodies at the north end were mined from stopes known as the Johnson, McCauley, East, Stevens, Messenger, Cameron, Twohy, Hampshire, Staisy, and Erwin stopes. The McCauley stope, the largest, was followed for a vertical distance of about 70 feet and at its greatest horizontal extent measured approximately 40 by 50 feet. The open cut has an outline measuring roughly 80 by 150 feet. The south-end workings are less extensive and include about 800 feet of crosscuts on two levels, in addition to an open cut, stopes, and a shaft 109 feet deep. None of these workings are now accessible.

The larger ore bodies are mined by square-setting and partial filling. The smaller stopes are supported by stulls, and at least one ore body was mined by shrinkage stoping. The square-set method allows rough sorting underground and is better adapted to mining the large irregular-shaped ore bodies and for supporting the slickensided rocks near them. After rough sorting underground the ore is trammed to outside bins and from them drawn to sorting tables, where the low-grade material is discarded and the ore of shipping grade is dropped to loading bins.

GEOLOGY

The ore at the Queen of Bronze mine occurs as disconnected bodies, irregular in outline, and ranging in size from mere stringers to deposits containing as much as 10,000 tons. The ore minerals do not form a solid body within the limits of the deposits but are interspersed with bands and irregular areas of altered wall rock, much after the manner of a mineralized shear zone. The more persistent bodies strike approximately east, some trending a little north of east and some a little south of east. Dips vary greatly, but the rake of the deposits, with exception of the Hampshire ore body, is to the south. The Hampshire ore body (fig. 23) rakes to the west. The deposits everywhere show effects of intense postmineral faulting. Slickensided surfaces and brecciated rock are conspicuous within the ore bodies and in the enclosing rocks, but are much less evident away from the mineralized areas. Drag effects and crushing are pronounced near faults in some of the more tabular deposits. (See pl. 16, A.)

The ore is enclosed in greenstone, including both metabasalt and metagabbro, and is found near the contacts with serpentine but was not observed within serpentine. Next to the ore the original characteristics of the greenstones are usually obliterated by processes which have changed the rocks to a mass of chlorite, quartz, and calcite containing disseminated sulphides. Very little sericite was observed in thin sections.

Oxidation extends as much as 100 feet below the surface, but sulphide minerals prevail below 50 feet. Surface processes have produced high-grade oxidized ore near the surface, and sulphide enrichment undoubtedly was an important process at shallow depths, though it has not contributed greatly to the copper content below a depth of 100 feet.

Before sorting, the ore in some of the stopes, as indicated by samples taken by G. E. Stowell, mining engineer, has a metal content

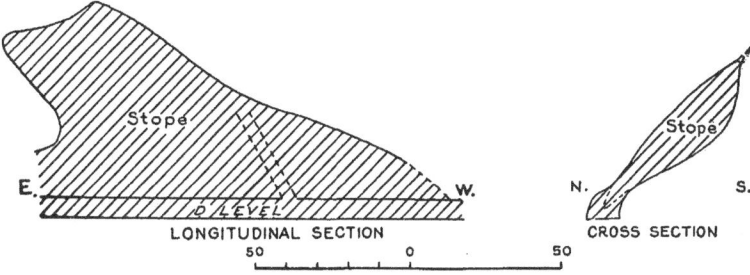

FIGURE 23.—Sections through Hampshire stope on D level of Queen of Bronze mine

of 4 to 7 per cent of copper and 0.04 to 0.1 ounce of gold to the ton. The average copper content of the oxidized ores near the surface was considerably higher and according to Kay [82] was over 10 per cent. A sample taken by Mr. Stowell across 5 feet of the ore in the East stope assayed 4.8 per cent of copper and 0.04 ounce of gold to the ton. Another sample from the same stope taken across 7.8 feet assayed 4.0 per cent of copper and 0.10 ounce of gold to the ton. A sample across 6 feet in the top of the east end of the same stope assayed 4.7 per cent of copper and 0.08 ounce of gold, and a sample across 7.5 feet on the sill floor directly below assayed 7.0 per cent of copper and 0.06 ounce of gold. A sample across 12 feet of ore in the south drift of the 50 level assayed 5.4 per cent of copper and 0.06 ounce of gold. A "bunch" of ore 3 feet wide from the top of the stope above the 70 level assayed 6.4 per cent of copper. If these assays represent the average metal content of the ore in place, it follows that considerable waste is readily eliminated by sorting, as the shipping ore in the past has averaged about 8.3 per cent of copper and about 0.13 ounce of gold and 0.16 ounce of silver to the ton.

[82] Diller, J. S., and Kay, G. F. Mineral resources of the Grants Pass quadrangle and bordering districts, Oreg.: U. S. Geol. Survey Bull. 380, p. 79, 1909.

The mineralogy of the ore is comparatively simple. The hypogene sulphide minerals include pyrite, chalcopyrite, and, in much smaller amounts, pyrrhotite and sphalerite. The proportion of each varies considerably from place to place, and as a result rough sorting is practiced underground to eliminate the material high in pyrite and low in chalcopyrite. Textural relations indicate that two generations of pyrite are present. Older massive pyrite is cut by well-defined fractures containing chalcopyrite, whereas the younger pyrite occurs as disseminations or as grains in more or less parallel arrangement along irregular fractures in chalcopyrite, as shown in plate 17, *A*. Supergene chalcocite is present near the surface, where it replaces hypogene sulphides. The supergene replacement has been selective, chalcopyrite being replaced to a much greater extent than pyrite. The abundant oxidation products include malachite, azurite, cuprite, iron oxides, and chrysocolla; tenorite and native copper have also been reported.[33] The sulphides follow fractures in the quartz, whereas the calcite cuts both quartz and sulphides and, in places, is abundant enough to deplete the grade of the ore seriously. (See pl. 17, *B*.)

ECONOMIC CONSIDERATIONS

The largest ore shoot had a vertical extent of about 70 feet, but most of the stopes are less than 50 feet high. The horizontal outlines of the ore bodies are irregular and vary greatly at different altitudes. Projections of the ore bodies as a basis for tonnage estimates of ore in place are therefore hazardous and not reliable.

Many faults cut and displace the ore bodies, and it seems likely that the numerous disconnected shoots represent segments of larger and more continuous deposits. Gouge-filled faults and stringers containing quartz, calcite, and sulphides become increasingly numerous toward the larger ore bodies and if used in conjunction with wall-rock alteration should serve as a guide toward mineralized areas.

Ore bodies probably exist in some of the unexplored areas. Recent prospecting has proved the presence of good-sized deposits as deep as the D (lower) level, and the type of mineralization does not indicate that other ore bodies are not present, even below this level. Winchell [34] states that "the apparent relation of the ore bodies to the present erosion surface suggests that they owe their final position to the work of downward percolating surface waters." This is interpreted to assume that the present position of the ore bodies is largely

[33] Diller, J. S., and Kay, G. F., op. cit., p. 77.
[34] Winchell, A. N., Geology and mineral resources of Jackson and Josephine Counties, Oreg.: Mineral Resources of Oregon, vol. 1, No. 5, pp. 73, 257, Oregon Bur. Mines and Geology, 1914.

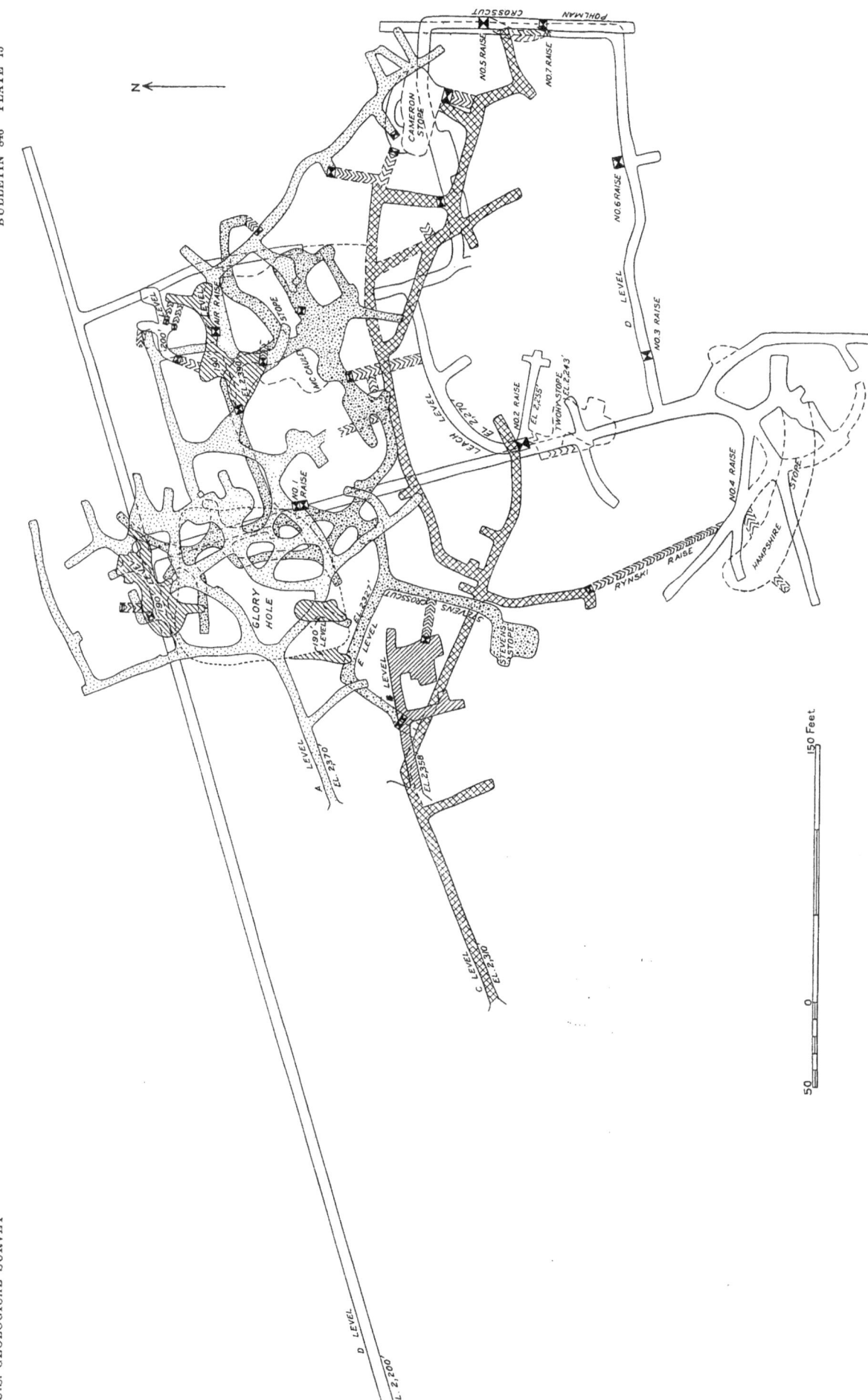

PLAN OF NORTH-END WORKINGS OF QUEEN OF BRONZE MINE.

A. TABULAR BODY OF SULPHIDE ORE (*s*) TERMINATED BY A FAULT ON LEACH LEVEL,
QUEEN OF BRONZE MINE.

g, Greenstone.

B. BOULDER ORE FROM COWBOY MINE.

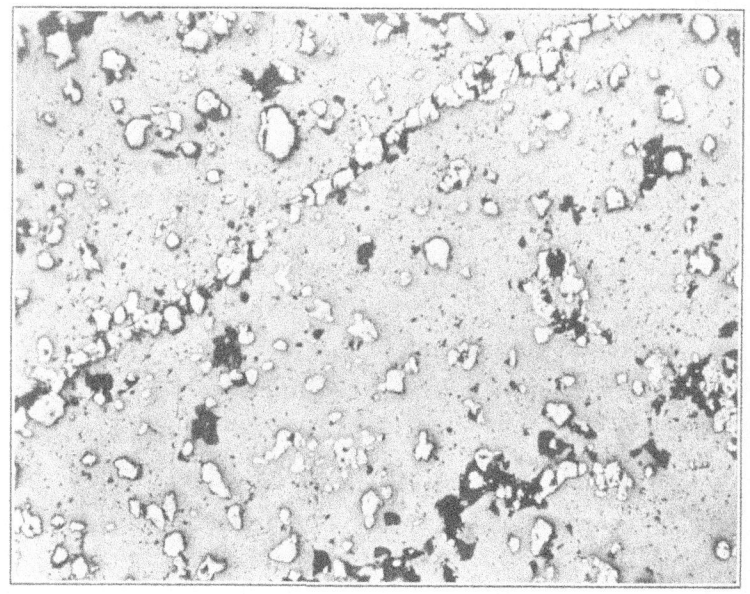

A. PYRITE (WHITE) IN ORIENTED ALINEMENT AND SCATTERED THROUGH
CHALCOPYRITE (LIGHT GRAY).

Black areas are holes.   Polished section, parallel light, enlarged 100 diameters.

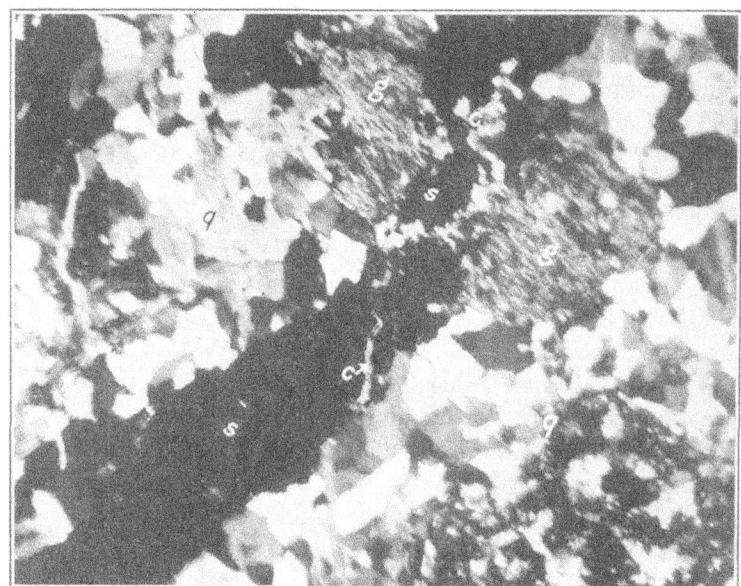

B. SULPHIDE VEIN (*s*) CUTTING QUARTZ (*q*) AND FRAGMENTS OF CHLORITIZED
GREENSTONE (*g*) AND IN TURN CUT BY VEINLETS OF CALCITE (*c*).

Thin section, crossed nicols, enlarged 28 diameters.

PHOTOMICROGRAPHS OF ORE FROM QUEEN OF BRONZE MINE.

due to sulphide enrichment. Sulphide enrichment has undoubtedly been an important process near the surface, but evidence of it is lacking in the deposits below a depth of 100 feet. The wall-rock alteration, the presence of pyrite, chalcopyrite, and pyrrhotite as the principal ore minerals, and the absence of chalcocite or other supergene (secondary) copper minerals in the deeper ore bodies confirm the hypogene origin of these deposits. Therefore little change in the mineral composition or metal content of the ore can be expected for at least several hundred feet below the zone of enrichment.

WALDO MINE

The Waldo mine is in the SW. ¼ sec. 36, T. 40 S., R. 8 W., about a mile east of Takilma and half a mile south of the Queen of Bronze mine. The property is developed through several tunnels, which are now inaccessible. The property, under option to the Queen of Bronze Mining Co. in 1930, is owned by the Waldo Smelting & Mining Co. No complete production records of the Waldo mine are known to be available, although considerable of its ore was treated at the Takilma smelter, and it is said that shipments from this mine were made through Grants Pass after the smelter ceased operating.

The ore is found near the contact of serpentine and greenstone. Near the mine a roof pendant of greenstone, about 500 feet across, is surrounded on all sides by serpentine, and numerous smaller isolated blocks of greenstone occur in the serpentine near the contact. According to Winchell [35] the ore on one of the levels had a vaguely veinlike form (fig. 24), trending about N. 60° E. and dipping about 45° SE., whereas in a stope 30 feet above the strike was apparently N. 70° W. and the dip 50° SW. At a higher level the ore occurred in a well-defined fissure striking N. 75° E. and dipping 55° SE. The ore on the dump more nearly resembles that of the Cowboy mine than the Queen of Bronze ore. It is commonly rounded and enclosed in slickensided gouge. Chalcopyrite is the principal ore mineral. No cubanite was found in the samples collected. Pyrrhotite tends to follow incipient fractures and is more abundant than at the Queen of Bronze, but pyrite is less abun-

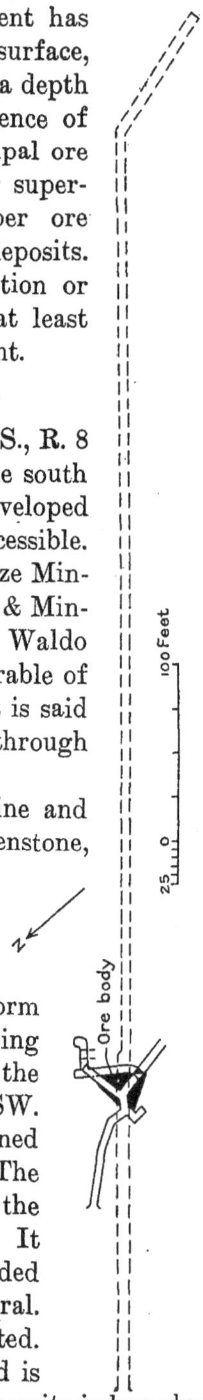

FIGURE 24.—Plan of a part of the underground workings of the Waldo mine. (After A. N. Winchell)

100 Feet

25 0

Ore body

[35] Winchell, A. N., op. cit., p. 253.
165255—33——3

dant. Malachite, azurite, and iron oxides are the most common oxidation products.

LILLY MINE

The Lilly mine is in the SE. ¼ sec. 35, T. 40 S., R. 8 W., about 2 miles by road southeast of Takilma. The property, consisting of eight claims, was located in 1897 by Kameel Khoeery, of Takilma, who sold it to the Knob Hill Mining Co. in 1910. It was relocated by Mr. Khoeery in 1911, and a half interest was sold to Mrs. Charlotte Johnson, also of Takilma, in 1915. According to Mr. Khoeery the mine has produced about 300 tons of copper ore which contained from 16 to 28 per cent of copper. Part of the ore was mined from open cuts and part from underground openings. A tunnel several hundred feet long and with several crosscuts has been driven in the direction of the open cuts.

The ore occurs as irregular bodies in greenstone rocks that are completely surrounded by serpentine. The ore mined from the open cut was partly oxidized, but only hypogene sulphides have been found in the lower tunnel. The ore from the lower tunnel is much like the ore from the Cowboy mine and is composed principally of chalcopyrite, cubanite (copper-iron sulphide), pyrrhotite, and sphalerite in a gangue of altered greenstone, quartz, and calcite. Quartz has preceded the sulphide minerals. Chalcopyrite and cubanite occur as irregular masses and as lathlike intergrowths. Pyrrhotite in irregular veinlets cuts the intergrowths of cubanite and chalcopyrite and is therefore considered to be somewhat younger. Sphalerite is present in small amounts, but its textural relationship was not determined.

LYTTLE MINE

The Lyttle mine is on the north side of Page Creek in the SW. ¼ sec. 1, T. 41 S., R. 8 W., at an altitude of about 2,700 feet. By road the property is about 3 miles southeast of Takilma, and in an air line it is a mile directly south of the Waldo mine.

Most of the ore was mined through a glory hole, now about 110 feet long by 80 feet wide and 20 feet deep. Maps show that an upper tunnel, 160 feet long, was driven under the glory hole. At approximately 100 feet from the portal an inclined winze was sunk which is said to be in ore at a depth of 60 feet. These workings are now caved and inaccessible. A lower tunnel, at a considerable distance below the upper workings, is 570 feet long. From it a raise, 53 feet long, has been run toward the bottom of the 60-foot winze on the level above but has not connected.

The Lyttle mine is said to have produced 1,500 tons of ore, part of which was smelted at Takilma. The property is now owned by the Queen of Bronze Mining Co.

The ore is found in greenstone not far from a serpentine contact. The greenstone in the vicinity of the mine is completely surrounded by serpentine, and much serpentine was cut in the lower tunnel. The ore deposit appears to be similar to those at the Queen of Bronze mine, although the mineralogy more nearly resembles that of the Cowboy ore. Chalcopyrite, cubanite, and pyrrhotite are the principal sulphide minerals. Malachite, azurite, and iron oxides can be observed in the open pit and in the dumps.

ORIGIN OF THE DEPOSITS IN GREENSTONE

The evidence indicates that the deposits in greenstone were formed in and along fracture zones by hot ascending solutions. The presence of abundant chlorite, fine-grained quartz, and disseminated hypogene sulphides in the greenstone wall rocks enclosing the ore affords evidence of alterations by hot mineral-bearing solutions during the process of ore deposition. Pyrrhotite, found in all the mines, and cubanite, found in some of them, indicate that the deposits were formed at high temperatures and at considerable depth. Schwartz [36] has concluded, on the basis of laboratory experiments, that deposits in which the intergrowth of chalcopyrite and cubanite is found were formed above 400° C. and probably above 450° C.

Structurally, the deposits are related to fractures that were developed in greenstones much after the fashion of shear zones. Mineralizing solutions, high in silica and magnesia, ascended through the fractures to form first principally chlorite and quartz. After fracturing, hypogene sulphides were introduced, for the most part in areas that had previously undergone silicification. Another period of fracturing followed the deposition of the sulphides, and the fractures were healed by calcite, which also replaced the rock outward from the fractures. This calcite was introduced late in the history of the ore deposits and in some of the mines is plainly visible cutting faults that offset the ore. Oxidation and sulphide enrichment are the processes that have more recently prevailed. They have, however, been active only near the surface and have taken no part in the formation of most of the ore. Winchell's view [37] that the position and form of the ore bodies seem to be due to the work of meteoric (surface) waters is not substantiated by available evidence. Because of postmineral movements, the original outlines of the deposits are uncertain. Faulting has obliterated their original shape, but in some of the mines, particularly the Queen of Bronze, several of the segments have marked tabular characteristics.

[36] Schwartz, G. M., Intergrowths of chalcopyrite and cubanite—experimental proof of the origin of intergrowths and their bearing on the geologic thermometer: Econ. Geology, vol. 22, No. 1, pp. 44–61, 1927.
[37] Winchell, A. N., op. cit., p. 73.

COPPER DEPOSITS IN SERPENTINE

COWBOY MINE

LOCATION AND DEVELOPMENT

The Cowboy mine (pl. 18) is in the NE. ¼ sec. 11, T. 44 S., R. 8 W., 3 miles by road southeast of Takilma, at an altitude of about 2,600 feet on a steep slope overlooking Page Creek. A serviceable road has been built as far as the lower tunnel, but ore from the upper and principal workings must be hauled on sleds for about an eighth of a mile over a rough course.

The ore body is developed through tunnels and by various raises, stopes, and winzes. About 2,000 feet of tunnels have been driven— 350 feet on the upper level, 200 feet on the intermediate level, 500 feet on level 2, about 100 feet at the East Cowboy, and, largely during 1930, 850 feet on the lower (No. 3) tunnel. Tunnel 3 was driven with the hope of intersecting the west ore body 200 feet below the present stopes, but at the time of the writer's visit, in August, 1930, it had not reached its objective.

HISTORY AND PRODUCTION

According to E. H. Messenger, the superintendent, a Mr. Strong discovered ore on what is now known as the East Cowboy about 1900 and excavated ore from an open pit. However, little work was done prior to 1903, when C. L. Tutt and associates, of Colorado Springs, Colo., purchased the property. The Queen of Bronze Mining Co., the present owner, acquired the mine in 1916 and, although the property has been leased at various times, has mined most of the ore produced. From 1916 to 1919 a total of 842 tons of ore was mined, and it is reported that ore was treated at the Queen of Bronze smelter between 1906 and 1910. However, most of the production is credited to the period from 1928 to 1930, when 75 cars were shipped. Mr. Messenger estimates a total production of about 100 cars, or, roughly, 5,000 tons. The value of the total production is estimated at $300,000.

GEOLOGY

The ore bodies at the Cowboy mine are found near the contact of greenstone and serpentine. The prevailing greenstones in the vicinity of the ore bodies are even-grained and fragmental varieties of metabasalt and medium-grained metagabbro. A highly altered greenish rock with large white phenocrysts, tentatively classed as metadiorite, has recently been exposed in the lower tunnel. Numerous masses of greenstone are included in the serpentine near the contact, and many of them have been found underground. The serpentine is normally dark green and, in general, has a high luster. Near the ore bodies, however, it contains much calcite and has a stony appearance resembling that of the altered greenstones.

The ore occurs along a fault zone in serpentine as a series of slightly curved lenslike bodies separated and surrounded by dark grayish-green gougy material consisting principally of fine-grained, felted antigorite (serpentine). The fault zone, in places from 6 to 8 feet wide, extends to the north and south beyond the ore limits. The lenses of ore are composed of rounded lumps of massive sulphide minerals or serpentine lumps with sulphide stringers, but, although the ore as mined resembles blocks of serpentine, it is easily distinguished by its greater weight. (See pl. 16, *B*.) In general, the ore lenses strike north and dip 45°–65° E. The angle of dip has increased with depth. The maximum length of the series of ore lenses is about 170 feet, and the thickness ranges from that of thin stringers to 7 or 8 feet. Oxidation and enrichment have occurred to a noteworthy extent only near the surface. An increase in copper content in the ore mined from the open pit was undoubtedly caused by sulphide enrichment, but the process has not added materially to the copper content below a depth of 50 feet.

MINERALOGY

The abundant hypogene sulphides are cobaltite, chalcopyrite, cubanite, sphalerite, and pyrrhotite. Chalcocite occurs as a supergene sulphide, and malachite, cuprite, tenorite, hematite, and limonite are the more common oxidation products. In order of abundance the gangue minerals are serpentine, calcite, quartz, and epidote. With the exception of serpentine the gangue minerals are not readily visible in hand specimens, although postsulphide calcite is in some places evident along fracture surfaces. The microscope shows, however, that calcite constitutes a considerable part of the ore and of the wall rocks next to the ore.

The cobaltite resembles pyrite in hardness and crystal outline but differs from it in color. Although microchemical tests reveal considerable iron in the cobaltite, the crystal form and lack of anisotropism distinguish it from glaucodot (a cobalt-iron-arsenic sulphide). Cubanite and chalcopyrite differ considerably in color and degree of anisotropism. Pyrrhotite resembles cubanite but is readily distinguished from it in polished sections by a greater relief. Sphalerite is fairly abundant and is readily distinguished by its gray color. In addition, it almost everywhere contains oriented blebs of chalcopyrite or pyrrhotite.

The succession in the deposition of the sulphide minerals is the normal one as defined by Lindgren despite the fact that repetition occurs. There appears, however, to be a reversal in the succession of the gangue minerals. According to Lindgren[38] the normal order of mineral deposition in deposits of this general class is silicates, quartz, carbonates and other gangue minerals, cobaltite, pyrrhotite, sphalerite, and chalcopyrite. The order of formation as actually determined is serpentine, calcite, epidote, quartz, cobaltite, sphalerite, chalcopyrite and cubanite, pyrrhotite, sphalerite, and calcite. The succession was thrice interrupted by fracturing—once after the deposition of the gangue minerals, again after the formation of the cobaltite, and again after the deposition of the sulphides but before the deposition of the later calcite. Serpentine was formed before the deposition of the older calcite, because veinlets of this calcite clearly cut felted plates of the serpentine, thus illustrating that serpentinization had taken place, in part at least, before the deposition of the ore. Epidote appears to have formed after the older calcite, possibly in part from the reaction of hydrothermal solutions upon it. Quartz veinlets clearly cut the epidote, and veinlets of sulphides, in turn, cut all three of these gangue minerals. (See pl. 20.) The sulphides have replaced calcite more readily than the other gangue minerals, and in most places this differential replacement of calcite is very noticeable. Cobaltite was the first sulphide mineral deposited. A period of fracturing followed, and then later sulphides were introduced, for the most part along the fractures. (See pl. 19, A.) Sphalerite is the first sulphide known to have formed after the cobaltite. If other sulphides preceded the sphalerite the evidence of them in the ores studied has been completely destroyed. Chalcopyrite and cubanite formed after the sphalerite and, where associated, they occur as bladelike intergrowths. Of the two, chalcopyrite is considerably more abundant. Pyrrhotite succeeded the cubanite and chalcopyrite. It occurs as irregular masses, as veinlets in or along grains of older minerals, as lentils in chalcopyrite and cubanite, and as oriented blebs and laths in sphalerite. The blebs and laths were certainly formed by replacement along cleavage directions in the sphalerite, as they occur only where sphalerite is known to be replaced by pyrrhotite. (See pl. 19, B.) At other places oriented blebs of chalcopyrite are numerous, but blebs and laths of pyrrhotite are missing. Lentils of pyrrhotite cut intergrowths of cubanite and chalcopyrite at various angles and in some places are parallel to the intergrowths.

[38] Lindgren, Waldemar, Magmas, dikes, and veins: Am. Inst. Min. Met. Eng. Trans., vol. 74, p. 88, 1927.

to serpentine contacts, is so general that Diller [45] felt that the serpentine had much to do in producing the ore deposits, although he points out that the serpentine itself rarely contains bodies of ore except copper.

The shape of the original ore bodies is not easily interpreted. Tiny sulphide stringers and disseminations are found in the wall rocks next to the more massive sulphide ore, and stringers usually extend for some distance beyond the termination of the ore lenses. It is therefore believed that the original ore bodies were roughly lens-shaped but that the lens shape has been accentuated by postmineral movements. The strongest postmineral movements were probably an accompaniment of the general deformation of the region, although the processes attending serpentinization, which is essentially a hydration process producing a considerable increase in rock volume, may have contributed to the stresses causing the movements, at least locally. Movements resulting from stresses, whatever their origin, normally cause adjustments along numerous irregular fractures in serpentines, but where harder rocks are included in the serpentine the adjustments would tend to follow fractures passing around the more resistant bodies. Well-defined, slickensided fractures of this type can be observed in the Cowboy mine next to the ore bodies and around greenstone inclusions. Attempts have been made to follow these fractures away from the ore bodies, particularly the well-defined hanging-wall fracture on the No. 2 and intermediate levels, but without success, owing to the fact that the fracture tends to lose its identity a short distance from the ore.

The principal events in the genesis of the ore at the Cowboy mine may be outlined as follows: After the peridotite rocks had become solid and while these rocks were deeply buried, fractures or lines of weakness developed parallel to the greenstone-peridotite contact. Ore-bearing solutions, derived either from the parent magma or from a younger intrusive body, in places forced their way along the fractures, or lines of weakness, and deposited gangue and ore minerals. Calcite appears to have been introduced first, followed by epidote, which may have developed partly by the reaction of the hot solutions with the introduced calcite. Quartz was next introduced. After the deposition of the quartz, stresses within the rocks caused fracturing, and the fractures controlled in a large measure the deposition of the sulphides that followed. Cobaltite was the first sulphide introduced. It was fractured, and the later sulphides were introduced along the fractures. Sphalerite was the first sulphide to

[45] Diller, J. S., op, cit. (Bull. 546), p. 20.

chalcopyrite and pyrrhotite as the principal ore minerals and, as described, appear to resemble the Cowboy deposit very closely. In all the deposits described by Hershey and by Butler and Mitchell little or no quartz or calcite is reported. Butler and Mitchell apparently believe that the ore minerals in the "boulder" deposits described by them were originally distributed throughout the igneous rocks but have been segregated in the positions now found during the changes accompanying the serpentinization of the containing rocks.[42] For the deposits in northern California Hershey [43] says:

Perhaps the molten rock came into contact with and absorbed rocks containing ordinary copper deposits, thus deriving an unusual copper constituent which was widely disseminated in certain portions of the peridotite and related basic rocks but during serpentinization became segregated with the iron minerals. However, it remains an open question as to whether the segregation was connected with the solidification of the magmas or with the subsequent serpentinization.

The presence of cobaltite, pyrrhotite, and chalcopyrite in a basic igneous rock and the apparent scarcity of quartz, calcite, or other gangue minerals characteristic of veins is at once suggestive of a deposit formed by magmatic segregation. The mineral association undoubtedly indicates that the ore was formed at high temperatures and at considerable depth, but the fact that the sulphides have been introduced into calcite, epidote, and quartz, which, as shown by the microscope, are abundant, points to another mode of origin—that is, to a hydrothermal deposit originating under conditions of high temperature and at considerable depth but in and along fractures. According to Schwartz [44] the presence of chalcopyrite and cubanite intergrowths indicates a temperature of formation above 400° C. and probably above 450° C.

The mineral assemblage in the Cowboy ore, in the light of present knowledge, points quite definitely to a deep-seated origin. The source of the ore minerals, however, can only be surmised. Granitic rocks have been intruded into the serpentine in areas closely adjacent to Takilma and no doubt are not far beneath the surface in the Takilma vicinity, although none were found at the surface. These later granitic rocks are believed by most investigators to be the source of many ore deposits in southwestern Oregon, particularly of the gold-quartz veins. However, the occurrence of copper deposits in very close association with serpentine, or in greenstone at or close

[42] Butler, G. M., and Mitchell, G. J., op. cit.

[43] Hershey, O. W., op. cit., p. 430.

[44] Schwartz, G. M., Intergrowths of chalcopyrite and cubanite—experimental proof of the origin of intergrowths and their bearings on geologic thermometers : Econ. Geology, vol. 22, No. 1, p. 60, 1927.

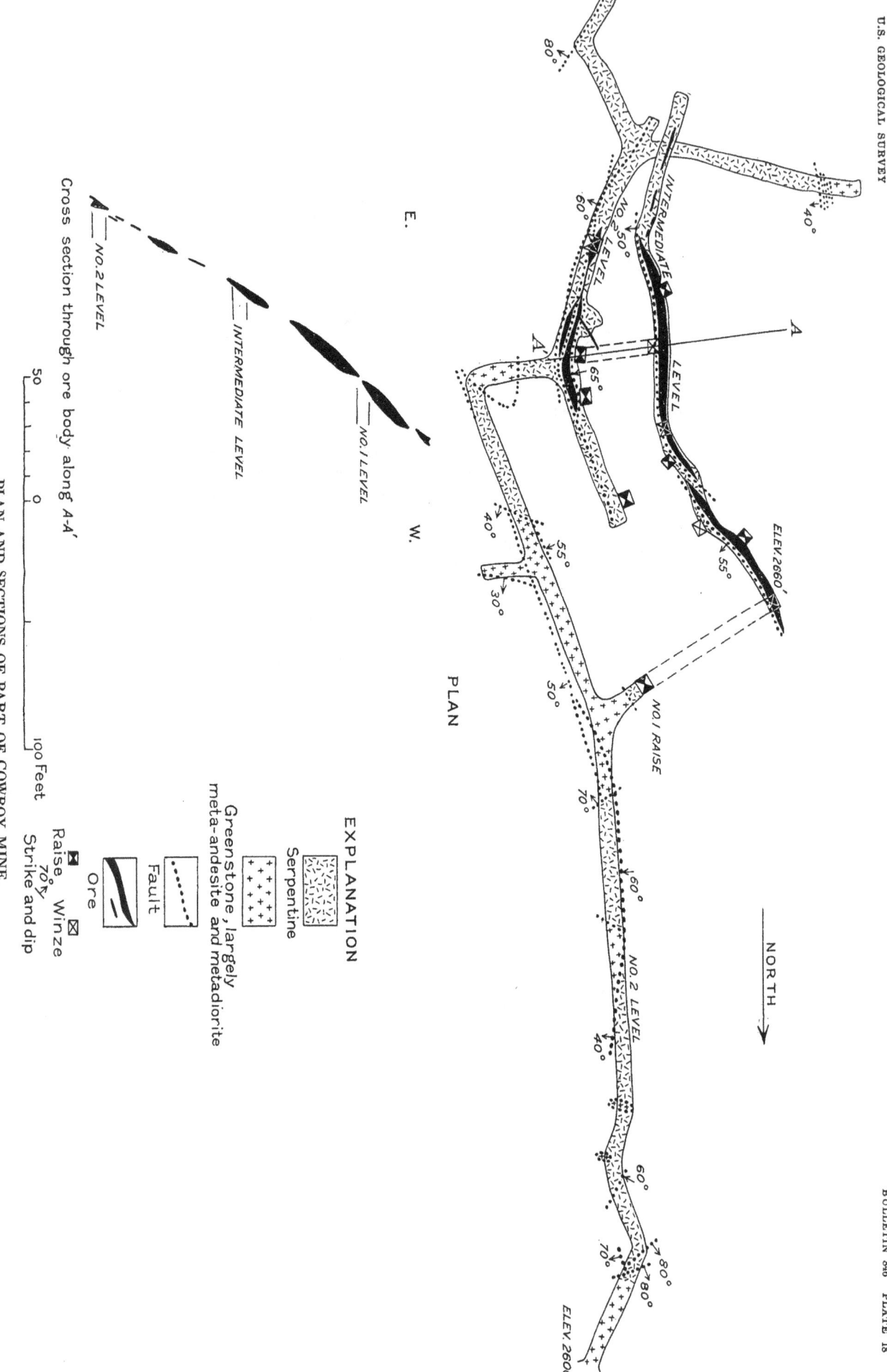

PLAN

EXPLANATION

Serpentine

Greenstone, largely
meta-andesite and metadiorite

Fault

Ore

Raise   Winze
70°
Strike and dip

NORTH

Cross section through ore body along A-A'

NO.1 LEVEL

INTERMEDIATE LEVEL

NO.2 LEVEL

E.

W.

PLAN AND SECTIONS OF PART OF COWBOY MINE.

50   0   100 Feet

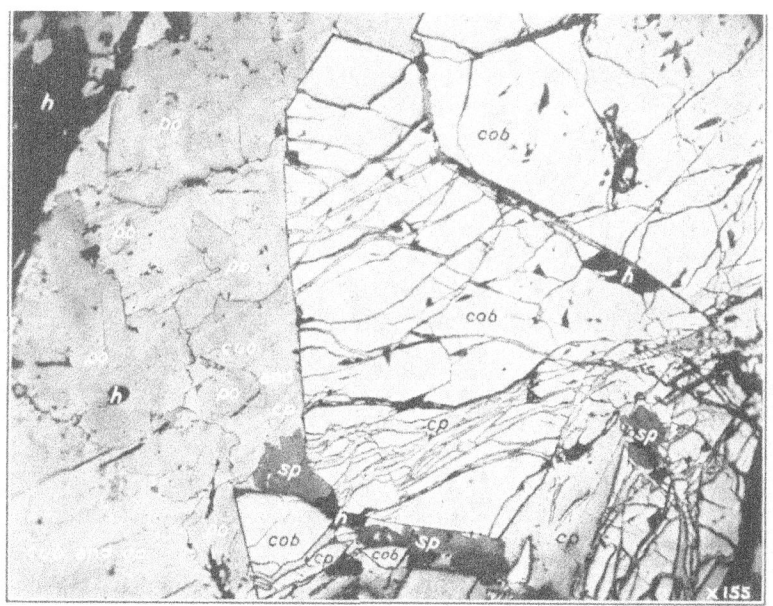

A. CHALCOPYRITE *(cp)* AND SPHALERITE *(sp)* ALONG FRACTURES IN COBALTITE *(cob)*; PYRRHOTITE *(po)* CUTTING INTERGROWTHS OF CHALCOPYRITE AND CUBANITE *(cub)*.

h, Holes.   Polished section, parallel light.

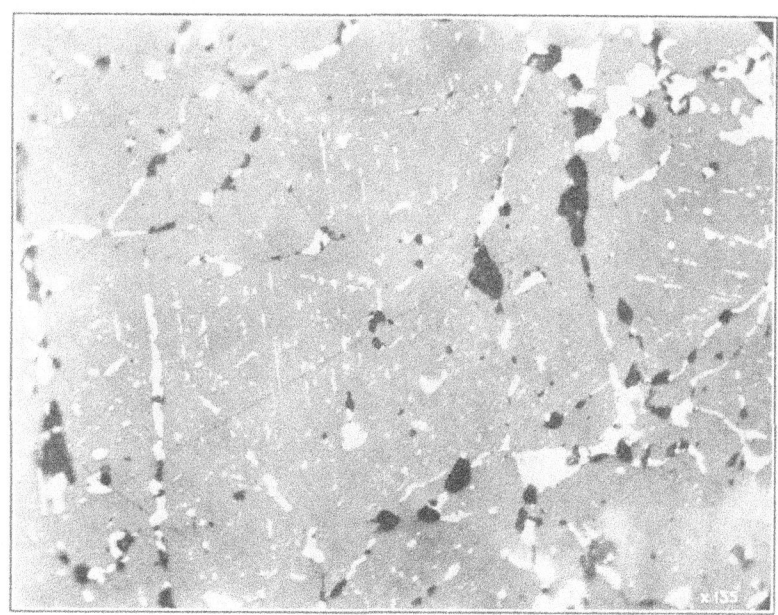

B. SPHALERITE (GRAY) CONTAINING PYRRHOTITE (WHITE) ALONG VEINLETS AND AS TINY BLEBS ORIENTED IN CLEAVAGE DIRECTIONS.

Black spots are holes.   Polished section, parallel light.

PHOTOMICROGRAPHS OF ORE FROM COWBOY MINE.

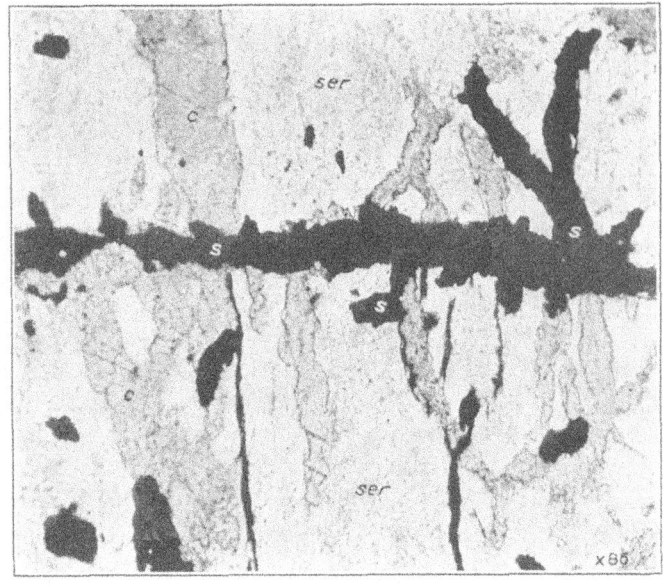

A. SULPHIDE VEINLET (*s*) CUTTING VEINLETS OF OLDER CALCITE (*c*) IN SERPEN-
TINE (*ser*).

Thin section, parallel light.

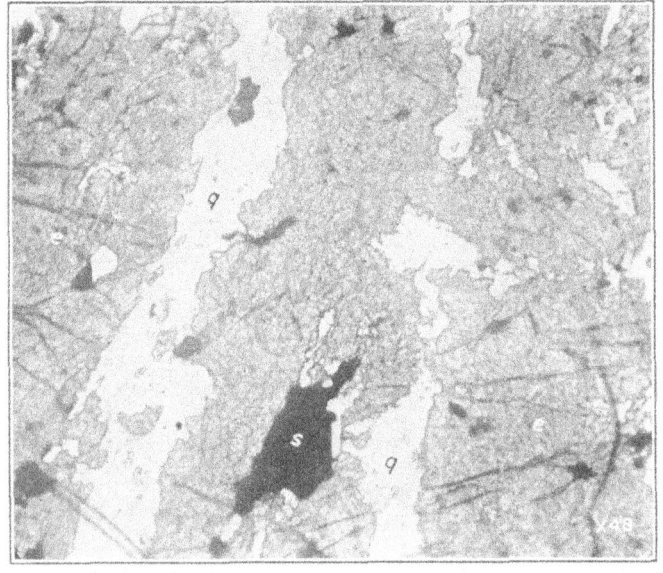

B. QUARTZ VEINLETS (*q*) CUTTING EPIDOTE (*e*).

s, Sulphides. Thin section, parallel light.

PHOTOMICROGRAPHS OF ORE FROM COWBOY MINE.

to serpentine contacts, is so general that Diller[45] felt that the serpentine had much to do in producing the ore deposits, although he points out that the serpentine itself rarely contains bodies of ore except copper.

The shape of the original ore bodies is not easily interpreted. Tiny sulphide stringers and disseminations are found in the wall rocks next to the more massive sulphide ore, and stringers usually extend for some distance beyond the termination of the ore lenses. It is therefore believed that the original ore bodies were roughly lens-shaped but that the lens shape has been accentuated by postmineral movements. The strongest postmineral movements were probably an accompaniment of the general deformation of the region, although the processes attending serpentinization, which is essentially a hydration process producing a considerable increase in rock volume, may have contributed to the stresses causing the movements, at least locally. Movements resulting from stresses, whatever their origin, normally cause adjustments along numerous irregular fractures in serpentines, but where harder rocks are included in the serpentine the adjustments would tend to follow fractures passing around the more resistant bodies. Well-defined, slickensided fractures of this type can be observed in the Cowboy mine next to the ore bodies and around greenstone inclusions. Attempts have been made to follow these fractures away from the ore bodies, particularly the well-defined hanging-wall fracture on the No. 2 and intermediate levels, but without success, owing to the fact that the fracture tends to lose its identity a short distance from the ore.

The principal events in the genesis of the ore at the Cowboy mine may be outlined as follows: After the peridotite rocks had become solid and while these rocks were deeply buried, fractures or lines of weakness developed parallel to the greenstone-peridotite contact. Ore-bearing solutions, derived either from the parent magma or from a younger intrusive body, in places forced their way along the fractures, or lines of weakness, and deposited gangue and ore minerals. Calcite appears to have been introduced first, followed by epidote, which may have developed partly by the reaction of the hot solutions with the introduced calcite. Quartz was next introduced. After the deposition of the quartz, stresses within the rocks caused fracturing, and the fractures controlled in a large measure the deposition of the sulphides that followed. Cobaltite was the first sulphide introduced. It was fractured, and the later sulphides were introduced along the fractures. Sphalerite was the first sulphide to

[45] Diller, J. S., op, cit. (Bull. 546), p. 20.

form after the cobaltite. It was followed without interruption by chalcopyrite, cubanite, pyrrhotite, and sphalerite. The cubanite was probably deposited as a solid solution in chalcopyrite but separated out during cooling as bladelike intergrowths when the proper temperature was reached. The occurrence, late in the series, of pyrrhotite followed by sphalerite indicates a recurrence of higher temperature before the succession was completed. Fracturing followed the deposition of the sulphides, and the younger calcite was introduced along the fractures. Eventually the deposits were exposed by erosion, which recently has kept pace fairly well with oxidation and enrichment, as there is very little evidence of either below a depth of 50 feet.

ECONOMIC CONSIDERATIONS

At the Cowboy mine, as at the Queen of Bronze, very little reserve ore is blocked out, owing in large measure to the nature of the deposit. Development and mining of necessity proceed simultaneously, because of the irregular outline of the ore bodies. In other words, it is necessary to mine the ore in order to delineate its outline. The present ore shoot has been followed downward on its dip for about 170 feet. Within this distance six major lenses were found, and it seems likely that with further prospecting downward others will be discovered. The ore body on level 2 is shorter horizontally than on the levels above, but there is no reason to suspect that the ore will end abruptly at this point. The feasibility of further prospecting down the dip, however, will depend largely upon the demand for copper. Thus far, prospecting has not revealed a series of lenses in horizontal alinement on level 2, despite the fact that the prospecting has been done along the fault in which the ore occurs. However, if the genesis of the deposit is correctly interpreted, there is no reason to believe that other ore lenses do not exist at the Cowboy mine in the unexplored ground in the immediate vicinity of the proved ore. Recent work on the East Cowboy has disclosed oxidized copper ore of good grade, and this deposit appears worthy of further prospecting when the copper market justifies the expenditure.

The mineralization at the Cowboy mine is of the deep-zone type, and as oxidation and enrichment have occurred only near the surface, the ore cannot be expected to differ greatly from that on level 2 for another several hundred feet, providing it should continue downward for that depth. Furthermore, the mineralogy and metal content of undiscovered ore bodies in the immediate vicinity, if they exist, should not differ greatly from those of the bodies already known.

FUTURE OUTLOOK FOR THE COPPER MINES

The copper deposits of the Takilma-Waldo district have been the most productive of the Oregon copper mines. They are credited with a gross production of about $1,700,000 and appear capable of producing a considerable future tonnage, but because of the high mining and transportation costs they can be worked at a profit only when the demand for copper is strong. Very little ore is actually blocked out in any of the mines, and yet some can be seen in most of the stopes. Largely because of the irregular outlines of the deposits and because the shoots terminate suddenly, mining and development proceed more or less simultaneously, and hence very little ore is proved in advance of mining. The better ore has probably been mined out of the known ore bodies, but if the origin of the deposits is interpreted correctly there is reason to believe that undiscovered ore bodies exist above the deepest workings in several of the mines and, in some of them, even deeper than any of the present workings. The actual depth to which the ore deposits will extend, however, is uncertain. Most of the deposits in the region are enclosed in greenstones near serpentine contacts, and it is therefore evident that the depth to which these deposits may be expected to continue depends a great deal upon the depth to which the greenstone extends. At some of the mines the greenstones exist as shallow roof pendants and therefore cannot be expected to extend far vertically. Other mines are enclosed in larger roof pendants, and still others are in greenstone bodies that have not been isolated by the intrusion of serpentine. The last-mentioned bodies can be expected to extend deeper, and hence the likelihood is greater that the ore will continue to a greater depth. For those deposits which may extend downward for some distance, very little change can be expected in the mineralogy of the primary ore for several hundred feet below the shallow zone of enrichment.

CHROMITE DEPOSITS

Small deposits of chromite are known in several places near Takilma, and development work has been done on some of them. All occur as irregular bodies in serpentine or as irregular-shaped surface boulders which clearly have been derived from serpentine areas. Chromite deposits have been mined north of French Flat, in the NW. ¼ sec. 22, T. 40 S., R. 8 W., and in the NE ¼ sec. 11, T. 41 S., R. 8 W., and irregular bodies of chromite were disclosed in the serpentine bedrock in pit 2 of the Llano de Oro mine (pl. 21, *A*). Several large boulders of chromite occur along the road between Takilma and the Queen of Bronze mine and in other places. The total production of chromite in the Takilma-Waldo district is estimated at about 1,000 tons, which has come almost entirely from the Esterly mine.

ESTERLY MINE

The Esterly mine is about 2 miles northwest of Takilma, in the NW. ¼ sec. 22, T. 40 S., R. 8 W. The ore has been mined from a trench about 50 feet long and from several open pits. The larger openings are now inaccessible because of water. The mine was worked in 1918 by George Barton and is said to have produced about 1,000 tons of ore.[46]

The ore occurs as irregular bodies in serpentine and, as indicated by the trend of the openings, has a strike of about N. 20° E. The ore is typically mottled green and black and is composed principally of chromite and serpentine. Because of the inaccessibility of the mine openings, very little field evidence is available regarding the origin of the chromite. Thin sections show veinlets of fibrous antigorite cutting the chromite, but this evidence in itself is not sufficient, especially in view of the more recent studies,[47] to conclude that all or most of the chromite at the Esterly mine is of early magmatic origin.

OWENS MINE

The Owens chromite mine is in the NW. ¼ sec. 11, T. 41 S., R. 8. W. It is reached by a trail connecting with the East Fork road at the Owens farm. The deposit is opened by a tunnel about 40 feet long. Only a few tons of ore have been mined. A small body of serpentine crops out at the mine and is surrounded on all sides by fine-grained greenstone. The chromite occurs as two small lens-shaped bodies, one near the face of the drift and one about 15 feet from the portal.

PLACER DEPOSITS

HISTORY AND PRODUCTION

Placer mining began in the region of Takilma in the spring of 1853, with the discovery of gold on Althouse Creek. At about the same time sailors are said to have abandoned a ship on the coast and traveled overland to the " Sailor Diggings," near Takilma.[48] Raymond[49] quotes Doctor Watkins, a physician of long residence in the community, as follows:

A ditch was dug some 15 miles long at a cost of some $75,000 or $80,000 to bring water to the rich placers of this vicinity and when fairly well under

[46] Esterly, G. M., personal communication.

[47] Sampson, Edward, May chromite crystallize late?: Econ. Geology, vol. 24, No. 6, pp. 632–641, 1929. Ross, C. S., Is chromite always a magmatic segregation product?: Idem, pp. 641–645. Fisher, L. W., Origin of chromite deposits: Idem, pp. 691–721.

[48] Winchell, A. N., op. cit., p. 241.

[49] Raymond, R. W., Mines and mining in the States and Territories west of the Rocky Mountains for 1869, p. 213, 1870.

way paid for itself the first year. It paid heavy dividends to its stockholders for 10 or 12 years, and many parties who live sumptuously every day owe their fortune to their connection with the Sailor Diggings Ditch Co.

An early-day mining town, Allentown, located near the mouth of Allen Gulch, flourished for several years. Names of some of its early inhabitants are still legible on tombstones in an old graveyard a few hundred feet southeast of the Platerica mine pit. With the exhaustion of the more readily mined and richer gravel, mining activity declined until 1878, when work was started on the Deep Gravel mine. After a short time production again gradually declined until 1901, when the introduction of hydraulic elevators made possible the profitable working of the gravel of French Flat by hydraulic methods. In recent years the placer production of the Takilma-Waldo district has come chiefly from the Llano de Oro mine. Considerable unworked areas have been prospected on Llano de Oro ground, and these are reported to carry gold and platinum in sufficient amounts to yield a profit for several years.

No figures are available on the production of the placers of the Takilma-Waldo district before 1878, although undoubtedly the richest ground was worked prior to that time. The known placer production since 1878 is, in round figures, about $1,000,000. In the absence of authentic records of the early-day production, a rough estimate of the minimum production may be based on the extent of the ground that was worked and the costs of operation at the time. In addition to stream alluvium at Waldo and in Allen, Sailor, and other gulches, the early miners worked the surface material on the adjacent slopes, particularly the areas on or adjoining the patches of Tertiary conglomerate. Second-growth timber including trees 50 to 60 years old covers most of this mined ground and fixes rather definitely the time elapsed since that mining period. All accounts agree that the gulch placers were rich, which in those days meant that with primitive methods such as rocking and ground sluicing each miner could produce at least $4 to $10 a day by uncovering 1 or 2 square yards of bedrock. On this basis $2 a square yard is assumed as a safe estimate of the minimum yield. The areas observed to have been worked over during the early period aggregate at least 300 acres, or 1,500,000 square yards, from which the gold produced is estimated to be at least $3,000,000. The total minimum production up to the present time is therefore estimated at $4,000,000.

MINING METHODS

The shovel and rocker of the early days, with a capacity of 3 to 5 cubic yards of gravel a day, have been supplanted in the Takilma-Waldo district by large-scale methods whereby hundreds of cubic yards of gravel is washed daily. Hydraulic giants are used entirely for mining. Water for their operation is brought in ditches for long distances, from the East Fork of the Illinois River. From one ditch a nozzle pressure of over 300 feet is obtained. Sufficient water is available for only about 7 months of each year, however, and hence the mines are worked continuously day and night during the mining season. The broken material loosened from the gravel banks by the giants is washed through sluice boxes and undercurrents (pl. 21, C), where the gold is collected with quicksilver or in a concentrate. In the more favorably situated mines the discarded gravel (tailings) is carried away by the natural run-off of the water, but in mines where there is but little grade it is lifted by hydraulic elevators (pl. 21, B) and distributed over the ground selected for tailings dumps.

CLASSIFICATION OF PLACERS

The placers are classified as Tertiary and Quaternary deposits. The Tertiary deposits are composed principally of cemented boulders and sandy material derived from the erosion of older rocks. In places the unweathered Tertiary formation contains sufficient gold to be minable, but in general the gold content is small. Weathering and leaching have, however, enriched the conglomerate near the surface and produced a mantle containing a relatively high gold content. This material, before it was mined out, covered a considerable area.. Most of this ground was mined 50 or 60 years ago and is said to have been very productive. The most valuable placer deposits remaining in the region are the Quaternary deposits, which have been derived in part through the reworking of the Tertiary conglomerate by erosional processes that have caused a reconcentration of the gold. Almost if not quite all the richest of the reworked placers are located on or below areas of Tertiary conglomerate where streams or rain wash have transported and re-sorted the material. (See pl. 11.) Valuable Quaternary deposits have thus been formed at the Llano de Oro and Deep Gravel mines and in several small gulches traversing or receiving wash from the Tertiary conglomerate—for example, in Sailor, Allen, Fry, and Scotch Gulches. The small gulches were worked out by the early-day miners, but considerable areas of the transported deposits remain unworked at the Llano de Oro and Deep Gravel mines, and with the mining methods now in use, a steady production may be expected to continue from them for a number of years.

TERTIARY PLACERS

GENERAL FEATURES

The Tertiary placer mines of the Takilma-Waldo district are all in the Tertiary conglomerate. This formation is found in the northwestern part of the district, but, owing to faulting and erosion, only remnants of it are now present. The remnants have a linear distribution over a distance of 4 miles. Although the origin of practically all the placer gold in the district can be traced to this formation, only a small portion of it has been mined in place. The early-day miners must have produced considerable gold by mining the residual surface mantle over the conglomerate, but no reliable estimates of this production are available. Exclusive of the mantle material and the re-sorted gravel, the Tertiary conglomerate has probably produced gold worth $100,000.

The Tertiary conglomerate is composed for the most part of large, highly altered boulders in a sandy matrix. Sandstone beds occur in the lower part of the formation but are scarce in the upper part. The formation is cut by many joints and some faults, and veinlets filled with quartz and calcite are fairly numerous. Several different types of bedrock underlie the conglomerate. The most southerly outcrops are underlain by greenstones; those in the northern part of the district are underlain principally by serpentine and Horsetown(?) sandstone. In general, the bedrock is regular in contour and very well suited for placer mining. The greenstone bedrock is colored purple and is greatly decomposed for several inches below the contact with the Tertiary conglomerate. A short distance beyond the limits of the overlying formation, however, the greenstones appear to be free from alteration. The intense alteration of the bedrock directly beneath the conglomerate formation is apparently connected with the intense alteration of the boulders within the formation. At some time past the rocks near the surface were apparently subjected to deep secular decay. The weathering agencies probably penetrated the porous Tertiary formation with relative ease, and in most places the underlying bedrock was attacked. A thick mantle of soil was thus formed on the bedrock beneath the conglomerate as well as beyond its limits, but, where this soft, altered material was not protected by the overlying beds, it was removed by erosion. Some of the overlying boulders may also have contained sufficient sulphide minerals to liberate considerable sulphuric acid when attacked by oxygenated surface waters. If so, this sulphuric acid could have attacked other boulders as well as the underlying bedrock.

Gold and platinum are distributed throughout the Tertiary conglomerate but probably are slightly more abundant near bedrock. Those more familiar with the mining conditions estimate that the gold content of the formation averages from 2½ to 3 cents a cubic yard. The gold is angular and flaky, and much of it is coated with a black film, apparently silica and iron oxide. Chromite, magnetite, limonite, hematite, ilmenite, epidote, zircon, and other heavy minerals occur in the concentrates with the gold and platinum. According to J. T. Logan, the ratio of platinum to gold is 1 to 75.[50]

Several ideas have been expressed regarding the origin of the gold and platinum, but, because it is difficult to prove them definitely, owing to the very small amount of these metals in a large volume of conglomerate, it is quite probable that differences of opinion will continue. Some operators believe that the gold has been derived from the quartz stringers which cut the Tertiary formation (pl. 22, B). If enough of the gold and platinum were derived from the stringers to make this source worthy of mention, the distribution should be related to the stringers. In other words, the highest gold and platinum content would be expected in or near the stringers, and a less amount away from them. This question was discussed with several of the operators familiar with the occurrence of the gold and platinum, and most of them stated that they had noticed no particular concentration of gold near the stringers. Furthermore, assays of the stringers did not show even a trace of gold or platinum. An acceptable explanation for the origin of the gold must take several conditions into account. (1) The gold and platinum are distributed throughout the formation; (2) there is apparently but little concentration along the bedrock; (3) most of the gold and platinum is flaky and angular in outline; (4) the boulders are softened by decay, and most of them are broken up during mining operations; (5) the gold appears to be more abundant where the boulders are most completely softened.

Under normal conditions, gold and platinum gravitate toward the bottom of the gravel as it moves downstream with the current, and as a result the richest ground in placer deposits is generally found at or near bedrock. As this is not true, or true only to a minor extent, of the Tertiary conglomerate, the gold and platinum must have remained suspended in the gravel as it moved downstream, or they must have been introduced after the gravel came into place, or they must have been enclosed in the boulders when the boulders were deposited, to be later liberated when the boulders disintegrated. Because the gold and platinum in the Tertiary conglomerate are not abnormally fine, it seems reasonable to assume that they should have gravitated downward during transportation, but this has happened only to a very slight extent, if at all. The objections to the hypothesis that the gold and platinum have been introduced after the deposition of the gravel have already been stated. The explanation that most of the gold and platinum were liberated by the disintegration of the boulders is best supported by the

[50] Hornor, R. R., Notes on the black-sand deposits of southern Oregon and northern California: U. S. Bur. Mines Tech. Paper 196, p. 31, 1918.

A. CHROMITE DEPOSIT (DARK) IN SERPENTINE IN NO. 2 PIT, LLANO DE ORO MINE.

B. HYDRAULIC ELEVATORS USED TO LIFT WATER AND TAILINGS FROM PLACER
PITS, LLANO DE ORO MINE.

C. UNDERCURRENTS FOR SAVING FINE GOLD AND PLATINUM, LLANO DE ORO MINE.

All photographs by G. M. Esterly.

A. LEACHED OUTCROP AT TURNER MINE.

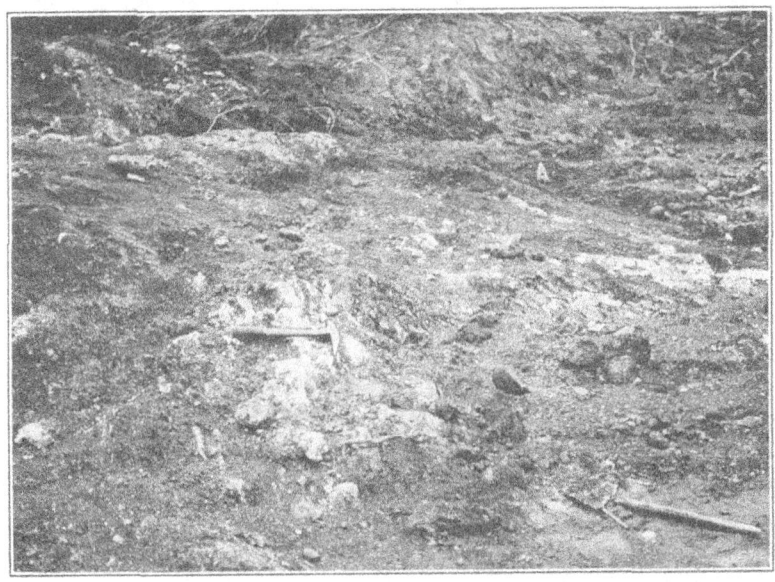

B. QUARTZ-ZEOLITE VEINLETS IN TERTIARY CONGLOMERATE, PLATERICA MINE.

Photograph by J. T. Pardee.

known conditions. The greenstones in southwestern Oregon are widely mineralized, and rich gold deposits have been found in them in many districts. Evidence of mineralization in the form of sulphides, principally pyrite, was seen in every thin section of greenstone studied, and in some of them sulphides were fairly abundant. It therefore does not seem unreasonable to expect the boulders of greenstone to carry a few cents in gold to the cubic yard. In addition, serpentine or other basic rocks are generally believed to be the most common source of platinum. The complete disintegration of the serpentine rocks in the Tertiary formation could easily account for the small amount of platinum present. The liberation of these metals, which is well accomplished in the Tertiary conglomerate, seems to be the critical factor. The relationship between the degree of alteration and the amount of gold and platinum available appears to be borne out by the observations of some of the operators that the highest metal content is found in areas of most complete rock decay.

HIGH GRAVEL MINE

The High Gravel (Osgood) mine is at the head of Allen Gulch, in secs. 33 and 34, T. 40 S., R. 8 W., near the drainage divide between the East and West Forks of the Illinois River, and is owned by F. H. Osgood, of Seattle, Wash. The mine includes several pits covering an area of approximately 150,000 square yards. Water for mining is taken from the East Fork some distance south of the Oregon boundary and is brought to the cuts through the Osgood ditch. The tailings are removed by natural run-off. The mine has been worked at different times by W. J. Logan, C. D. Cameron, an English syndicate, and others. Logan and Cameron leased the property during the period 1912–1917 and in the first 3 years took out $13,700 and in the last 2 years $2,000. Mr. Cameron [51] estimated the total production of all the cuts of the High Gravel mine, excluding the old workings along the bottoms of Allen and Scotch Gulches, at about $90,000.

The gold at the High Gravel mine is found in the Tertiary conglomerate, which is well exposed in several different banks and is composed mostly of poorly sorted boulders in a sandy matrix. Bedding is not plainly visible except in the lower part of the formation. The lower beds are sandy and have a purplish tint; the upper part of the formation exposed in the cuts is tan-colored and composed principally of large, poorly sorted boulders and sandy material. Distinct joints and veinlets occur throughout the formation. The conglomerate rests upon greenstone bedrock in several places.

[51] Cameron, C. D., personal communication.

(See pl. 13, *A*.) In the most westerly cut the contact strikes about N. 10° E. and dips about 20° E. At the High Gravel mine, as elsewhere, the conglomerate is composed of highly altered yet firmly cemented boulders of various types. Because of the induration attempts have been made to loosen the banks with explosives before hydraulicking, but according to reports this proved too costly for economical mining. The bedrock has a purplish tint and is highly decomposed wherever it is exposed beneath the conglomerate. It is cut by numerous fractures and small veins.

According to Mr. Cameron, the gold is distributed throughout the Tertiary conglomerate but is more abundant near the surface, where the formation is exposed to weathering. Much of it is coated with black material which makes amalgamation difficult. Mr. Cameron estimates the average gold content in the Osgood pits at about 3 cents a cubic yard.

CAMERON MINE

The pit here called the Cameron mine is near the head of Scotch Gulch, in the SW. ¼ sec. 34, T. 40 S., R. 8 W. It is owned by F. H. Osgood, of Seattle, Wash., but has been worked principally by lessees, chiefly J. T. Logan, C. D. Cameron, C. H. White, E. N. Bayse, and C. P. Johnson. A pit roughly 400 by 500 feet has been excavated by hydraulic giants. Water for the operation of the giants is supplied by the Osgood ditch, which takes water from the East Fork of the Illinois River south of the Oregon-California boundary. The tailings are removed by natural run-off. Most of the mining was done during the period 1924–27 although some gold was produced prior to 1909. The total production is estimated at about $9,000—$1,500 before 1909 and $7,500 during the period 1924–1927.[52]

The gold occurs in Tertiary conglomerate. As elsewhere, the lower beds are sandy and dark purple, and the upper exposed beds are light tan and consist principally of large, well-indurated boulders. Bedrock is not exposed beneath the conglomerate at the Cameron mine, but at the south side of the pit greenstones of the bedrock series are in fault contact with it. The fault that has dropped the conglomerate into contact with the greenstone strikes east and dips about 65° N., whereas the bedding in the conglomerate strikes N. 10° E. and dips 14° W. Boulders of greenstone, argillite, a talcky-appearing rock that is probably decomposed serpentine, and granitic rocks are most abundant in the conglomerate. The boulders are all well rounded and, for the most part, are highly decomposed. Even the granitic rocks readily fall to pieces when broken from their

---

[52] Cameron, C. D., and White, C. H., personal communication.

matrix. The matrix is principally sandstone, but the deposit is sufficiently indurated to make hydraulic mining difficult.

The gold is flat and flaky, and because much of it is covered with a black coating, amalgamation is difficult. According to C. H. White, the gold is distributed throughout the Tertiary beds but appears to be more abundant in areas of intense alteration. Mr. White estimates that the Tertiary formation in the Cameron mine contains on an average from 2½ to 3 cents in gold to the cubic yard.

PLATERICA MINE

During the winter and spring of 1929–30 a small cut was excavated by the Platerica Mining Co. in the NW. ¼ sec. 34 and the SW. ¼ sec. 27, T. 40 S., R. 8 W., near the head of Allen Gulch, about a mile west of Takilma. The gravel is mined by hydraulic giants, and the tailings are disposed of by natural run-off. Shallow workings southeast of the Platerica pit were excavated by early-day miners.

The Platerica Mining Co. has worked the lower beds of the Tertiary conglomerate, which where exposed in the pits is composed largely of greatly altered, rounded boulders in a sandy matrix. Numerous joints and some faults cut the conglomerate. Near the center of the cut an east-west normal fault, dipping about 45° N., has dropped the conglomerate into contact with the greenstone bedrock. The bedrock has a purplish tint, is greatly decomposed, and is traversed by numerous veins filled with quartz, epistilbite, and calcite. An assay of the material from the veins showed they contained no gold. According to J. L. Eggers, superintendent of operations, the gold occurs throughout the Tertiary conglomerate but is more abundant near bedrock. Here, as elsewhere, a large percentage of the gold is coated with black material. Much of it is collected in a black-sand concentrate along with chromite, magnetite, hematite, platinum, and other heavy minerals. It is reported that the hematite contains gold, although none was seen in polished sections of it. Mr. Eggers states that the conglomerate above bedrock averages about 2½ cents in metallic content to the cubic yard.

QUATERNARY PLACERS

Quaternary deposits have produced most of the gold and platinum in the Takilma district. Nearly all the valuable deposits have formed just below outcrops of Tertiary conglomerate, and the highest gold and platinum contents have been found closest to or on the conglomerate. These relationships are so consistent that there is little doubt as to the principal source of these metals in the Quaternary deposits. In general, there are three different types of Quaternary placers—(1) those formed on slopes below outcrops of Tertiary conglomerate where the gold is associated with very little gravel,

(2) deposits in narrow gulches traversing Tertiary conglomerate, and (3) transported deposits to which the Tertiary and other formations have contributed material. Deposits of the first group were largely worked out during the early days, so that little is known regarding their productivity. Good examples occur in the SE. ¼ sec. 21 and the SW. ¼ sec. 10, T. 40 S., R. 8 W., where the gold collected on serpentine bedrock. Deposits of the second type are well illustrated by the placers of Sailor and Allen Gulches. They were richly productive but, like the deposits of the first group, were worked by the early-day miners, and hence little is known of the amount of gold they produced. Most of the gold and platinum in recent years has come from the deposits of the third group, illustrated by the Llano de Oro and Deep Gravel mines and portions of Fry Gulch.

LLANO DE ORO MINE

The Llano de Oro mine, formerly the Logan, Simmons & Cameron mine, has for many years been the most productive gold-platinum placer in Oregon. The property includes over 3,000 acres of land in secs. 8, 9, 10, 15, 16, 21, 22, and 27, T. 40 S., R. 8 W., although practically all of the mining has been confined to the S. ½ sec. 15, the S. ½ sec. 22, and the N. ½ sec. 27. The property is operated by George M. Esterly, of Waldo.

The first important work on the Llano de Oro property was done south of the highway near the center of sec. 27 by early-day miners. C. H. White, who was acquainted with one of the miners, states that they mined gold worth $80,000 from this place. Later George Simmons, Frank Ennis, and Theodric Cameron took $110,000 out of Carroll Slough.[53] J. T. Logan mined the gravel on French Flat from 1907 to 1917, when the property was sold to G. M. Esterly. Mr. Esterly has worked the property almost continuously, during the mining seasons, up to the present time. He estimates the production in gold and platinum since 1917 at about $225,000 and the total production of the entire property at about $500,000.

Since 1907 most of the work at the Llano de Oro mine has been confined to the vicinity of French Flat. Four pits have been excavated, covering in all an area of over 30 acres. The depths of the pits vary considerably from place to place. For example, the depth to bedrock in pit 3 is about 8 feet on the west side and about 18 feet on the east side, whereas the average depth of the Logan or no. 1 pit is more than 30 feet, and at one place in it the tailings were elevated 50 feet. The company owns three ditches known as the upper, middle, and lower, together with three water rights to 500, 518, and 1,100

---

[53] Historical data furnished by C. D. Cameron and G. M. Esterly.

miner's inches from the East Fork of the Illinois River.[54] The total length of the ditches is over 15 miles. During the mining season, which averages about 7 months yearly, sufficient water is available to operate 2 giants in the pits, 2 hydraulic elevators, and 1 giant for stacking tailings. When the plant is operating steadily from 15,000 to 30,000 cubic yards of gravel, depending largely upon the seasonal water supply, is washed each month.

Both the Tertiary conglomerate and the Quaternary Llano de Oro formation have been worked at the Llano de Oro mine, but the latter has been by far the most productive. In only one place on Llano de Oro ground, in the SW. ¼ sec. 15, has the Tertiary formation been washed for its gold content. At this place the formation is well exposed in several cuts, where it can be seen resting upon serpentine in fault contact. The fault, which in part defines the eastern boundary of the Tertiary formation, strikes north and dips 65° W., whereas the normal contact dips 20° W.

The Llano de Oro formation consists of gravel, sand, and clay, is in general poorly sorted, and ranges in thickness from less than 1 foot near the edges to nearly 50 feet, but within the prospected areas on French Flat averages about 18 feet. Few boulders with diameters exceeding 6 inches are present. The bedrock varies at different localities. At several places it is Tertiary conglomerate; at other places serpentine or Horsetown (?) sandstone. The gold and platinum are concentrated near bedrock, although prospect holes show that some gold is distributed throughout most of the formation.

Most of the gold is angular and is associated with platinum chromite, magnetite, ilmenite, hematite, limonite, epidote, zircon, and other heavy minerals. Chromite was abundant enough in some of the areas of serpentine bedrock to be troublesome in the sluice boxes. The platinum occurs as flattened scales with rounded corners, which range in size from tiny grains to pieces over 2 millimeters in cross-section. Picked grains of platinum from the concentrate were analyzed by E. T. Erickson of the chemical laboratory of the United States Geological Survey, who reports that " the sample consists largely of platinum and ruthenium with smaller proportions of iridium and osmium. A small quantity of gold and slight quantities of palladium and rhodium were also detected." According to Mr. Esterly, platinum accounted for one tenth of the value of the clean-ups when it was worth $110 an ounce. In other words, the ratio of platinum to gold in the mined areas on French Flat is about 1 to 50.

In 1921 L. A. Levensaler, mining engineer in charge of prospecting for Mr. Esterly, estimated that the unmined gravel on French Flat

---

[54] Hornor, R. R., op. cit., p. 29.

within the prospected areas would average about 18 cents to the cubic yard. According to Mr. Levensaler, the value of the ground worked by J. T. Logan in the upper (No. 1) pit averaged 22½ cents a cubic yard, and that worked by Mr. Esterly at the other places in the same pit averaged 33½ cents a cubic yard. Kay [55] states that the gold content of the gravel mined in Carroll Slough was about 12½ cents a cubic yard.

DEEP GRAVEL MINE

The Deep Gravel mine is in Butcher Gulch, in secs. 16, 17, 20, and 21, T. 40 S., R. 8 W. Four deep pits covering a total area of approximately 50 acres and shallow pits covering well over 15 acres constitute the principal workings. The deep pits are designated, from north to south, Joe Smith Gulch, Wadleigh No. 2, Weimer, Wadleigh No. 1, and Johnson pits. The mine was first worked about 1874 by George and Walter Simmons. W. J. Weimer and sons purchased the property in 1878. In 1900 the ownership passed to the Deep Gravel Mining Co., in which Mr. Weimer retained an interest. In 1911 the Waldo Consolidated Mining Co. obtained an option on the property, but when the payments were not completed the ownership reverted to the Deep Gravel Mining Co. A. E. Reams, of Medford, Oreg., at present owns two thirds of the stock and acts as the representative of the company. Mr. Weimer stated that until 1908 about $130,000 had been expended on the property and it had produced $250,000.[56] Since 1907 the mine has produced about $26,316 in gold.[57] The Deep Gravel Mining Co. owns 350 acres of patented placer land, 410 acres of land held by mineral location, and a water right to take 2,800 inches of water from the East Fork of the Illinois River at a point a short distance west of Takilma.[58]

Most of the production of the Deep Gravel mine has come from the Llano de Oro formation, but recently Charles Johnson, of Takilma, excavated a small cut in Tertiary conglomerate in the S½ sec. 21. The Tertiary formation is here almost identical in appearance with the exposures at the Cameron mine, in Scotch Gulch. The lower beds are purplish conglomerate and sandstone; the upper beds are tan conglomerate composed of poorly sorted, coarse boulders which are fairly well indurated with sandy material. Like those at the Cameron mine, the boulders of the Tertiary conglomerate in the Johnson cut are for the most part highly decomposed. On the west they are in fault contact with Cretaceous sandstone.

[55] Diller, J. S., and Kay, G. F., Mineral resources of the Grants Pass quadrangle and bordering districts, Oreg.: U. S. Geol. Survey Bull. 380, p. 74, 1909.
[56] Kay, G. F., op. cit., p. 74.
[57] Data supplied by Victor C. Heikes, of the United States Bureau of Mines, and published with permission of owner.
[58] Hornor, R. R., op. cit., p. 32.

At the Deep Gravel mine, as elsewhere, the Llano de Oro formation is composed of gravel, sand, and clay and except in the lower 10 feet contains but few boulders over 6 inches in diameter. Stratification is well shown in some places. The thickness of the formation ranges from less than 1 foot near the edges to over 80 feet. Joe Johnson, of Takilma, assisted in the sinking of two prospect pits south of Mr. Potter's house. According to Mr. Johnson, the shafts passed through sand and clay containing lenses of fine gravel and at about 70 feet entered sandstone bedrock. A 2-foot layer of gravel on bedrock prospected very well, but above this layer the gold was sparsely distributed. So far as known, the bedrock in the various pits is either Cretaceous sandstone or Tertiary conglomerate. According to Kay[59] the bedrock in Joe Smith Gulch was 30 feet below the stream bed of the West Fork of the Illinois River, and hence hydraulic elevators were necessary to lift the gravel after the coarse gold had been removed on the riffles of a short sluice. After being elevated, the gravel was washed through another sluice 400 feet long in which the finer gold was collected. According to Kay[60] the average value of the pay gravel over a period of five years was about 25 cents to the cubic yard.

FRY GULCH

Fry Gulch is in secs. 28 and 33, T. 40 S., R. 8 W. Much of the gravel in it was worked in the early days, but some unworked ground remains. Two northward-trending branches of Fry Gulch join near the quarter corner between secs. 28 and 33. Both branches, as well as the main gulch for about 1,500 feet below the junction, have been mined. The east branch heads at the High Gravel mine, and the gold in it was clearly derived from the Tertiary conglomerate. The west branch heads near a flat summit close to the quarter corner of secs. 32 and 33. The boulders in it are similar to those in the east branch, but the source of the gold is not known, although it probably came from a patch of the Tertiary conglomerate, now completely eroded. Like Sailor Gulch and other small gulches receiving the wash from the Tertiary conglomerate, Fry Gulch was undoubtedly a rich placer, but, because much of the mining was done in the early days, no records of production are available.

In 1930 A. L. Bailey was working in a small cut near the mouth of the west branch. The gravel in the cut is composed of dark-red sand with pebbles of greenstone, serpentine, granitic rocks, sandstone, hematite, and chromite. The material is principally sand, and only a few of the boulders exceed 6 inches in diameter. Patches of unworked material of this sort extend up the west branch for about

[59] Kay, G. F., op. cit., p. 73.   [60] Idem, p. 74.

2,500 feet. The bedrock in Bailey's cut is Cretaceous sandstone, but in the east branch and farther up the west branch the gravel rests upon serpentine. According to J. L. Eggers, the production from about 1,650 cubic yards of gravel in Bailey's upper cut was $1,000, or about 60 cents a cubic yard.

RESERVES OF PLACER GRAVEL

Under present conditions it is doubtful whether the Tertiary conglomerate should be classified as among the reserve placer deposits. If, however, at some time exceptionally low costs should prevail, much of the formation might prove to be workable. Existing remnants within the area mapped aggregate a square mile or more and evidently contain many million cubic yards of material. The Llano de Oro gravel is to be regarded as the chief source of future placer production. Information given by the reports of trustworthy engineers and from other reliable sources indicates that areas of this formation aggregating several hundred acres contain enough gold to be profitably mined. In addition there is much ground that is probably gold-bearing, and the areas of known and probable value together aggregate at least 1,000 acres. The deposit ranges from less than a foot to 80 feet or more in depth, and its volume probably equals or exceeds that of the Tertiary. The largest remaining body adjoins the Llano de Oro and Logan (Carroll Slough) mines. Smaller areas remain in Butcher Gulch and in Fry Gulch below Waldo. In the prospected areas, the available information indicates that the gold content ranges generally from 10 to 60 cents a cubic yard, with streaks that are much richer.

BLUE CREEK DISTRICT

GEOLOGY

The geologic map of the Blue Creek district (fig. 25) includes about 4 square miles situated approximately 5 miles southwest of Takilma. Although most of the mining in this district has been done at the Turner (Albright) mine, the mapped area includes several other prospects.

The rocks of the area have been grouped into four units—the Galice formation, of Jurassic age; greenstones of Paleozoic or Mesozoic age; serpentine of Cretaceous age; and recent alluvium. The Galice formation, of sedimentary origin, occupies about one-third of the area mapped; greenstones and serpentine, of igneous origin, occupy most of the remainder. Small areas of recent alluvium occur along the larger streams.

The Galice formation near the Turner mine consists principally of sandstones (partly arkosic), slates and argillites, fine-grained conglomerates, and some interbedded greenstones. The conglomerates resemble those found near Takilma. The sandstones are firmly indurated, in most places massive, and prevailingly gray or reddish. The red color is probably due to surface weathering. The Blue Creek trail crosses a thick series of these red and gray sandstones immediately east of the quarter corner between secs. 10 and 11, T. 41 S., R. 9 W. The slates and argillites are fine-grained dark-gray to black rocks. They strike north of east and, for the most part, dip at steep angles to the southeast. The slates and argillites are

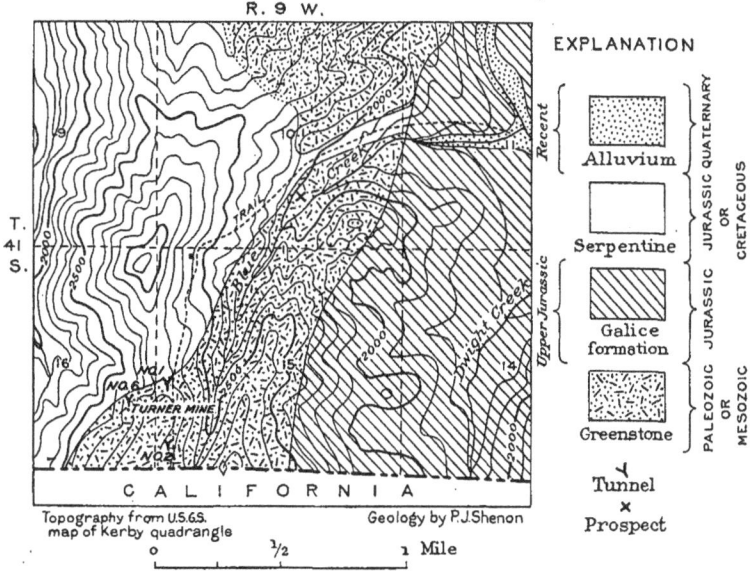

FIGURE 25.—Geologic map of the Blue Creek district

differentiated on the basis of cleavage. The slates split into thin laminae along planes independent of the original bedding, whereas the argillites are more massive. Upon fossil evidence Diller [61] has assigned the rocks of the Galice formation to the Jurassic period.

The greenstones consist of fine and medium grained rocks of prevailing dark grayish-green color, shown by the microscope and chemical analyses to be principally highly altered basalt and gabbro, which, because of intense alteration, have in this report been termed metagabbro and metabasalt. None are coarse-grained, and all show evidence of recrystallization.

The metagabbro is a medium grained granular rock in which the dark minerals appear, in hand specimens, to be much more abundant than the light minerals. The microscope, however, shows them to be

---

[61] Diller, J. S., Mineral resources of southwestern Oregon : U. S. Geol. Survey Bull. 546, p. 17, 1914.

about equally divided. All the minerals are greatly altered, the feldspars enough so to make their exact determination difficult. The plagioclase feldspar in the sections studied is everywhere partly altered to a very fine grained saussuritic product. Orthoclase, if present, occurs in minor amounts. Augite is largely altered to hornblende, biotite, and chlorite. The rock is very similar in appearance to the metagabbro of the Takilma district and, like that rock, is believed to have been intruded into the finer-grained greenstones and subsequently altered with them by dynamothermal processes.

The metabasalt is very fine grained, is dark grayish green to almost black, and near the ore bodies commonly has a greasy appearance. In thin sections the less altered portions of the metabasalts have pronounced basaltic textures, whereas the more altered areas are granular aggregates containing principally feldspar, epidote, chlorite, and saussuritic material too fine grained to classify. Some small patches of partly altered augite are usually visible. The green color of these rocks is due largely to the presence of chlorite and epidote. The evidence at hand indicates that the metabasalts were for the most part formed as flows.

Serpentine is the most widespread of the rocks near the Turner mine. As elsewhere in southwestern Oregon, it is dark green to almost black and is cut by many fractures, most of them with slick, greasy-appearing surfaces. In this vicinity, as in many other places, the serpentinization of the original rock is nearly complete. The texture and remnants of olivine and bastite (altered enstatite) indicate that the original rock was a peridotite composed largely of these two minerals. The serpentine, as at Takilma, is believed to have been intruded during late Jurassic or early Cretaceous time.

Considerable recent gravel and alluvium has been deposited along Elk Creek and its larger tributaries, Blue and Dwight Creeks. The gravel is composed of sand and pebbles derived from the erosion of the consolidated rocks of the region and includes various greenstones, slate, quartzite, argillite, chert, sandstone, several coarse-grained igneous rocks, and serpentine.

TURNER (ALBRIGHT) MINE

The Turner or Albright mine is just north of the California line, 45 miles southwest of Grants Pass, and 2½ miles by trail from the Redwood Highway. Between the highway and the mine the trail gains 1,200 feet in altitude. Waters Creek, the nearest railroad point, is 35 miles to the northeast. The property was located about 35 years ago and now belongs to Edward Turner and James Albright. It includes, according to Mr. Turner, three claims in sec. 15 and 260 acres of patented ground in sec. 16, T. 41 S., R. 8 W. Nine tunnels with numerous crosscuts have been driven which, in all, have a total length of over 3,000 feet. (See fig. 26.) No production has been reported.

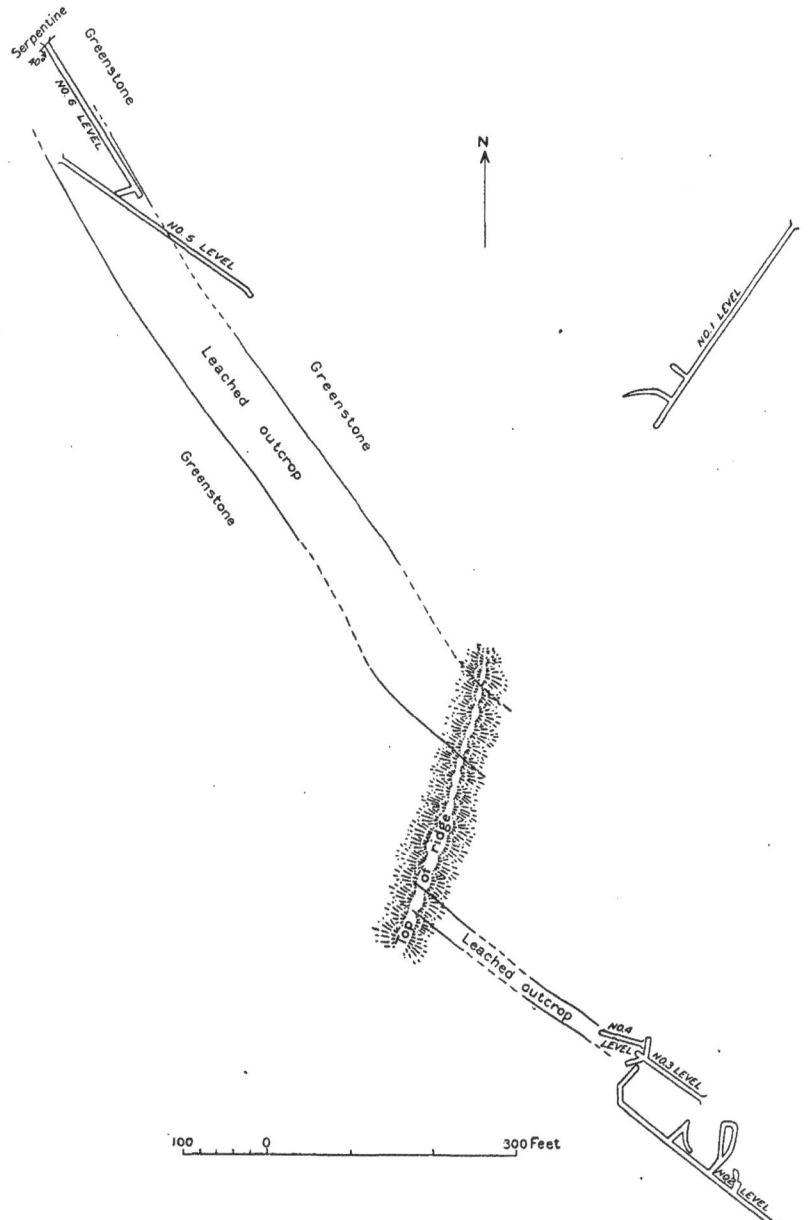

FIGURE 26.—Map showing relation of part of underground workings to leached outcrops at Turner mine

Two large bodies of porous iron-stained rock or " gossans ", enclosed in fine-grained greenstone, crop out at the Turner mine. One is about 80 feet wide and can be traced on the surface for 900 feet. The other averages about 20 feet in width and is well defined on the surface for over 300 feet. (See pl. 22, A.) Both gossans crop out prominently, but the narrower one is much more conspicuous because of the fact that it rises 30 to 50 feet above its surroundings. The larger gossan is partly prospected by tunnels 5 and 6. Both tunnels are near the surface and run through soft brown oxidized material and iron-stained greenstone. Some pyrite occurs near the face of tunnel 5, but the oxidation is elsewhere nearly complete. The smaller gossan is composed of porous brown, highly silicified material, which in places contains cores of unoxidized pyrite. In other places practically all of the iron has been removed, and there remains a cavernous white residuum composed principally of silica ribs. However, because of the abundant silica, a prominent outcrop has been maintained in spite of the thorough leaching. Beautiful specimens of the type of gossan described by Locke [62] as " botryoidal jaspery limonite " have been mined from one of the workings known as the " picture rock " tunnel. Sulphides are exposed in several tunnels beneath the smaller gossan. Of these, pyrite is by far the most abundant, although considerable chalcopyrite is associated with it in tunnels 2 and 3. In spite of the fact that the development work has thus far shown a high proportion of pyrite in the sulphide ore, the presence of considerable chalcopyrite with the pyrite at the face of tunnel 3, and below in tunnel 2, seems to justify more exploration on these levels. Because silicification makes the rock hard to mine by hand methods, work was stopped in the tunnels in the two places appearing most favorable for prospecting. An extension of tunnel 3 another 200 or 300 feet would add a great deal of information as to the probable worth of the property.

[62] Locke, Augustus, Leached outcrops as guides to copper ore, p. 138, Baltimore, Williams & Wilkins Co., 1926.

○

www.ingramcontent.com/pod-product-compliance
Lightning Source LLC
Chambersburg PA
CBHW081846170526
45167CB00007B/2916